Darwin's Legacy:

The Status of Evolutionary Archaeology in Argentina

Edited by

Marcelo Cardillo & Hernán Muscio

IMHICIHU

CONICET

Archaeopress Publishing Ltd
Gordon House
276 Banbury Road
Oxford OX2 7ED

www.archaeopress.com

ISBN 978 1 78491 270 3 (e-Pdf)
ISBN 978 1 78491 276 5

South American Archaeology Series No 24
Series Editor Andrés D. Izeta

Printed and bound in Great Britain by Marston Book Services Ltd, Oxfordshire

Contents

List of Figures

PREFACE

Hernán MUSCIO,[1] Marcelo CARDILLO[2]

[1] Consejo Nacional de Investigaciones Científicas y Técnicas (CONICET); Instituto Multidisciplinario de historia y Instituto de Arqueología (IDA). Universidad de Buenos Aires (UBA). hmuscio@gmail.com

[2] Consejo Nacional de Investigaciones Científicas y Técnicas (CONICET); Instituto Multidisciplinario de historia y Ciencias Humanas (IMHICIHU). Universidad de Buenos Aires (UBA).

This book collects the contributions to the symposium *"The current state of evolutionary archeology in Argentina"* that was held in Buenos Aires, for celebrating the 200th anniversary of Charles Darwin's birth and the 150th anniversary of the publication of *"On the Origin of Species"*. The meeting was sponsored by the IMHICIHU-CONICET (Instituto Multidisciplinario de Historia y Ciencias Humanas-Consejo Nacional de Investigaciones Científicas y Técnicas).

In Argentina, as Scheisohn (2009) noted, the early adherence to Darwin's ideas by scientists such as Florentino Ameghino influenced the subsequent reception of the Darwinian theory of evolution in archeology. But the development of this early intellectual evolutionary tradition was overshadowed by the discredit of Ameghino for his claims about the human origins in South America, opening the avenue for the subsequent dominance of two main paradigms, the culture-history school and the *Kulturkreise* archaeology (Boschín and Llamazares 1986).

The *Kulturkreise School* had the greatest impact on the archeology of Pampa and Patagonia, where it was established by the work of José Imbelloni, Osvaldo Menguín, and Marcelo Bórmida since the 30th and during the mid twentieth century. The *Kulturkreise* practitioners were devoted to the modal description of the archaeological record, particularly that of the lithic assemblages classified as specific areas industries or cultures . At the core of this paradigm were a hiperdiffusionist notion of cultural change and a radical idealism and antievolunism by which particular cultures were conceived as byproducts of the *ethos* of particular groups or races, expressed through the manufacture of artifacts and other behaviors, and dispersed from a few geographical centers.

On the other hand, during the mid-twentieth century, the culture-historical approach was developed mainly in Northwestern Argentina after the influential work of Alberto Rex Gonzalez. The research agenda of Gonzalez focused largely on the construction of local temporal sequences based on ceramic typologies and radiocarbon dating (the so called master sequences) (see González 1963, 1979). Although ignored by many of his disciples, a very important point was the call of Rex Gonzalez for a research strategy oriented to the study of cultural evolution. Nevertheless, the theoretical background of Rex Gonzalez was a blend between a strong typological thinking and the teleological explanations of cultural change that dominated the evolutionism of Julian Steward.

The path to the introduction of a scientific evolutionary framework in the archaeology of Argentina was opened in the late 70´s and the early 80´s by the application of the agenda of the New Archaeology, developed primarily in the United States with the work of Lewis Binford (1962) and in England by Graham Clark and his influential *Analytical Archaeology*. This perspective redirected almost all of the research goals by the adoption of a hypothetical-deductive model of science which resulted in the study of the adaptive dimensions of past human behaviors along the development of methodologies for detecting variability and patterning in the archaeological record. However, the evolutionism of the New Archaeology perpetuated the non-Darwinian explanation of change based mainly on cultural ecology and systemic functionalism. In this vein, the explanation of change was assumed to be always adaptive, following the orthogenetic notion of systemic adaptation by the continuous searching of homoeostatic equilibriums that characterized past adaptive systems. Therefore, under the influence of the ideas of Lewis Binford and Kent Flannery, adaptive systems became evolving units, and their reconstruction in archaeological terms was the main task of the so called systemic archaeology. Under this adaptationist framework the research on the relationships between past behaviors and past environments started as a programmatic goal through the study of past adaptive strategies.

In northwestern Argentina the work of Hugo Yacobacio and Daniel Olivera changed the focus from culture history towards the study of past human adaptations to high altitude environments (Yaccobacio et al. 1998,

Olivera 1988). Also, by virtue of this new agenda, the discussion of local evolutionary processes, such as the domestication of camelids in the Southern Andean region and the development of social complexity, was introduced.

In the archaeology of Pampa and Patagonia, the New Archaeology arouse as a vigorous reaction against the *Kulturkreise* archaeology and its bequest. Unlike the development of the new archeology in northwestern Argentina, in Patagonia, the introduction of this paradigm began with the discussion of processes that facilitated a gradual shift towards a Darwinian research agenda. After decades of anti-evolutionism thought, the archeology of Argentina began to be interested in processes like adaptive radiation and cultural divergence for explaining the archaeological variability of Patagonian record, which was dominated by the typological description of sites and lithic industries until then. In a series of seminal papers, Borrero (Borrero 1989, Borrero 1989-1990, Borrero 1992, Borrero 1993) settled the main distinctive issues of an early application of the evolutionary paradigm in Patagonia with an emphasis on biogeography, ecology, and taphonomy as the frameworks for explaining the variability of the archaeological record of Patagonia.

By the end of 80's the intellectual environment, especially at the University of Buenos Aires, began to be adverse to new archeology agenda, which had become the dominant paradigm of scientific archaeology by that time. This resulted in the proliferation of different lines of thought, among which Evolutionary Archaeology was highlighted by a large number of adherents that were working interactively in/for the development of a practical applicationto concrete case studies of a Darwinian evolutionary framework (Lanata 1995, 1996, Lanata and Borrero 1994, Scheinsohn 1997, 2002). Thus, throughout the 90's and well into the new millennium, a large number of students and young researchers broadened the application of the scientific evolutionism to the archeology of Patagonia (Barberena 2002, 2005, 2008) and northwestern Argentina (especially in the Puna, see Muscio 1999, 2004, Camino 2009, López 1998, López and Restifo 2009, Cardillo 2002, 2009), the northeast (Loponte et al 2006), and Cuyo and Sierras Centrales (Medina 2008, Neme 2009).

A peculiarity of the scientific evolutionism in Argentina archeology was the early adoption of a pluralist theoretical framework that integrated many different Darwinian approaches into a single evolutionary research agenda, thus ignoring the dicotomic opposition that characterized the development of the field in the U.S., between the so called selectionist school and other evolutionary approaches to human behavior, like Human Behavioral Ecology and the Theory of Cultural Transmission (see Scheinsohn 2008, 2009 for further discussion). Moreover, the English school also arrived at the beginnings of the new century and was added to this inclusive evolutionary paradigm (Shennan 2002). In the Argentinean tradition, an inclusive Darwinian theoretical framework evolved along a history of research where ecology and biogeography were always present by fuelling theory building around problems like the human colonization of the South-Andean region and Patagonia, and the emergence of productive economies in the puna region. Consequently, the intellectual challenge undertaken during the 90's was the articulation of biogeographical and ecological approaches with the key Darwinian sources of causality: variation and selective retention. In this way the search was one of integration into an inclusive paradigm of a variety of Darwinian approaches, a goal shared by a critical mass of researchers (Scheinsohn 2009).

Evolutionary archeology is the application of the Darwinian model of evolution to the study of the archaeological record -a fossil record- through the documentation of archaeological patterns of selective and non-selective evolutionary change. This theoretical framework in no way is reducible to the action of natural selection. Above all, Darwinian evolution implies a variational model of evolutionary change which is confronted to the transformational model of evolution based on typological thinking and teleological explanations. In this vein, the work of Borrero had the merit of challenge for the first time the omnipresence of the typological thinking in the archaeology of Argentina. This was a crucial step in the further development of the Evolutionary Archaeology in Argentina.

As Scheinsohn (2009) suggested, the reception and development of Evolutionary Archeology in Argentina comes from the influence of evolutionists like Ameghino and Holmberg. However, according to the author the acceptance and the variability of approaches framed within the broad spectrum of Darwinian approaches probably is the result of the relative scientific insularity of the academia in Argentina, regarding other centers of theory building, like the Anglo-Saxon countries, which are subject to other competitive forces in their academies. This volume is just a sample of the diversity of approaches and research issues vigorously conducted in Argentina.

INTRODUCTION: EVOLUTIONARY ARCHAEOLOGY A COMPREHENSIVE FRAMEWORK

Hernán MUSCIO,[1] Marcelo CARDILLO[2]

[1] Consejo Nacional de Investigaciones Científicas y Técnicas (CONICET); Instituto Multidisciplinario de historia y Instituto de Arqueología (IDA). Universidad de Buenos Aires (UBA). hmuscio@gmail.com
[2] Consejo Nacional de Investigaciones Científicas y Técnicas (CONICET); Instituto Multidisciplinario de historia y Ciencias Humanas (IMHICIHU). Universidad de Buenos Aires (UBA).

In the *Origins of Species* Charles Darwin concluded that the diversity of life arose by common descent through a branching pattern of evolution resulting from the accumulation of heritable variation preserved by natural selection. In the *Descent of Man* (1871) and by discussing the evolution of languages, Darwin asserted that cultural variants form genealogies by inheritance, and engage competitive relationships by which some variants proliferate whereas others become extinct. In this way Darwin clearly expanded the action of selection to culture. Thus, the contribution of Darwin to science was an inclusive and materialistic *model of evolutionary change* that encompasses everything which is subject to inheritance, variation, and differential replication.

The Darwinian model of evolutionary change is much more than evolution by natural selection as it describes the pattern of change that Lewontin (1983) called *variational evolution.* For Darwinian evolution, the occurrence of variation is a necessary condition, be it adaptive or neutral, blind or goal-oriented produced variation. Darwinian evolution implies that individuals in an evolving population vary from one to another, and that these variations may accumulate during evolution due to a sorting process by which some variants become either more or less common. The evolving populations may be formed by biological or cultural entities because both are subject to inheritance. Under the Darwinian model of evolution the sorting process -be it selection or random processes like drift- changes the proportions of the different variants in the population, figure 1-a. Conversely, Lamarckian evolution or "transformational evolution" does not assign any role to variation or sorting process. Transformational evolution is based on a typological concept of individuals by which all individuals of a population in a given environment simultaneously acquire the same structures and adaptations as the result of an inherent progressive tendency which drives them continuously towards greater complexity and to produce environmental induced hereditary adaptations (Figure 1). In this way Darwinian evolution, not the neo-Darwinism that emerged by the modern synthesis, does not preclude environmental induced variation as long this sort of variation became part of the variability subject to selection and other sorting forces (Kronfeldner 2007).

It is well know that cultural evolution has instances by which novel heritable variation arises as environmentally induced

behaviors by adaptive learning. But this variation remains subject to selection and other non-transformational population level process of evolutionary change, like imitation. The central issue is that problem solving cannot produce perfect solutions for adaptation, but more or less suitable alternatives. Then, the variation designed to solve a given problem will compete with other alternatives. Because of this, the evolution of culture is Darwinian, not neo Darwinian. This was proved by an increasing body of evidence that documents that culture mostly evolves following a pattern of descent with modification, i.e. the Darwinian branching pattern of evolution. As its biological counterpart, cultural evolution also has instances of reticulation by the horizontal transfer of information among independent lineages. Also, as in biology, the documentation of branching or reticulate patterns of cultural evolution is an empirical matter. Darwinian selection is a population level mechanism that sorts heritable variation (cultural or biological) by differences in the probabilities of competing variants to be successfully replicated. As a probabilistic mechanism, Darwinian selection is not reproductively based.

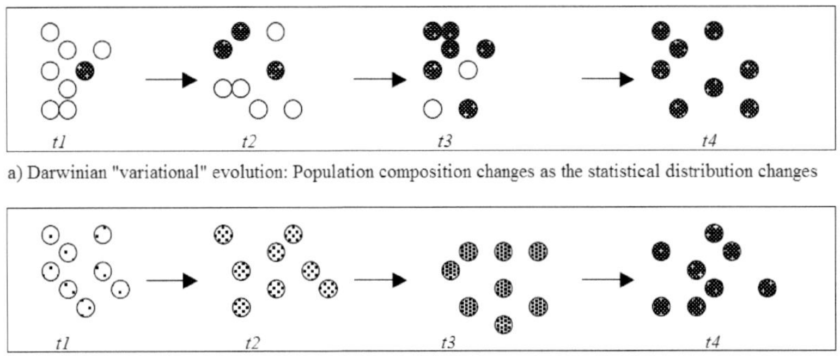

a) Darwinian "variational" evolution: Population composition changes as the statistical distribution changes

b) Lamarckian "transformational" evolution: Population changes becouse each individual changes in the direction of adaptation and evolution

FIGURE 1: VARIATIONAL VS. TRANSFORMATIONAL EVOLUTION, MODIFIED FROM KRONFELDNER 2007.

On this basis Evolutionary Archaeology is a scientific research framework that seeks to explain the archaeological record in terms of Darwinian evolution, through the documentation of archaeological patterns of selective and non-selective evolutionary change. To achieve this goal, Evolutionary Archaeology conceives the archaeological record as a fossil record that provides information about macroevolutionary patterns of evolution resulting from microevolutionary process of selection and transmission.

Differential reproductive success, by selection acting at the scale of the individuals, is one way by which alternative variants achieve replicative differentials. For instance a trait subject to cultural inheritance will result fixed in a population whenever it contributes more than its alternatives to the reproduction of the individuals possessing the trait. As an example, Rogers and Ehlrich (2008) found that 96 cultural traits of Polynesian canoes with effects on the survival and reproduction of the individuals evolved following a selective pattern of retention. Thus, this process acts through the increase or the decrease in the population of the individuals that transmit alternative cultural variants to others.

Another powerful way by which alternative variants achieve replicative differentials is *by selection acting at the level of the cultural variation,* through the existence of a bias acting on the replication of alternative cultural variants. For instance, with selection acting at the scale of the artifact, a trait will result fixed at the population level if it contributes more than its alternative variants to the replicative success of the artifacts possessing the trait. In this way, assume the existence of a replication bias for the quick removal of projectile points after impacting on preys. In this context, points without barbs will have greater replicative success than barbed points and because of this they will result fixed at the population level (Dunnell 2006, see also Hughes 1996). This process is called *cultural selection,* and acts through the differential replication of the cultural variation by processes of biased cultural transmission. As replicative success by cultural selection is decoupled of the differential reproduction of individuals, deleterious cultural variation can be circumstantially retained by maladaptive replicate biases until being removing by purifying selection acting against the individuals carrying the deleterious traits (Durham 1992).

Because natural selection has found in social learning a quick way to generate heritable adaptations, in cultural evolution it is expected that cultural selection prevails. However, the prevalence of some form of selection over another is an empirical question that should be discussed case by case.

Cultural transmission and traits replication depends on social learning. Based on the Dual Inheritance Model (Boyd and Richerson, 1985, Durham 1992), there is a bulk of mathematical work showing the way in which social learning produces microevolutionary change and cumulative evolution at the population level, as well as the reasons by which cultural transmission evolved by natural selection along human evolution.

In cultural evolution the emergence of novel variation may result from error transmission (the analogous of mutation) or from adaptive problem solving. This latter instance of problem solving invention is Lamarckian only in the restricted sense that introduces environmental instructed variation by adaptive learning; but because this variation depends on being fixed within the population of the action of selection (by imitation or by individual fitness gains), the population level pattern for the adoption of this variation is not transformational but Darwinian (see Ginóbili this volume).

In addition, because Darwinian evolution is variational, this framework also includes the evolutionary change that occurs by stochastic processes acting on cultural variation. Many cultural changes may take place on purely selectively neutral variation. Neutral evolution occurs by processes that randomly sort cultural transmission, so that some traits are differently replicated just by chance, for example through cultural drift. In small populations cultural drift might cause the extinction of cultural variants due to chance effects on transmission (Henrich 2004). On this basis, the theoretical distinction between selective and neutral traits is critical (Dunnell 2001).

Moroever, Darwin´s explanation of evolutionary change due to the action of selection preserving advantageous traits, replaced the essentialist logic for conceiving adaptation by a materialistic one. Since Darwin, adaptation is defined in reference to natural selection. In a historical sense, a trait is an adaptation if it has been produced by natural selection for its current function. This is the definition of adaptation as an outcome of selection. From a processual point of view, adaptation is a process that refers to the action of natural selection building somatic and behavioral features of the organisms that make them better suited to survive and reproduce. This last definition is the notion of adaptation as a process. As natural selection acting along the evolutionary past has built a flexible human phenotype, behavioral flexibility is an adaptation that gives rise to a diversity of adaptive traits, behaviors, and life histories as response to varying ecological contexts. Behavioral flexibility is part of what is known as phenotypic plasticity, the capacity of a single genotype to exhibit variable phenotypes in different environments. It is clear that plasticity is important because it can increase or decrease fitness, generate novelity, and facilitate evolution (West-Eberhard 2003).

But a blind trust in adaptive phenotypic flexibility leads to ignore the mismatches that may occur between human populations and their changing environments. In fact phenotypic plasticity often is maladaptive (Ghalambor et al, 2007). Darwin gave many examples of maladaptations in order to show the way in which species are ill fitted to their environments. Maladaptations may arise when the environment to which a species is adapted changes, or when species or populations disperse to unfamiliar environments. Muscio (1999) has showed the maladaptive pattern of faunal consumption during the colonization of the puna of Argentina. Importantly, a profound mismatch between a species and its environment leads to extinction by directional selection (Haldane 1957).

The archaeological application of the Human Behavioral Ecology research strategy analyzes the *adaptive design* of past behaviors through formal models based on the selectionist logic of fitness maximization which in turn, and for heuristic proposes, is usually modeled as an optimizing problem. This theoretical approach is based on the adaptive decision making that is thought to be part of the behavioral flexibility of humans. Under this approach a number of topics are of evolutionary interest, like the archaeological patterns of resource transfer and consumption, habitat selectivity and space use, technological change, and social interactions.

The Human Behavioral Ecology approach is a powerful theoretical framework for studying past social dynamics. For instance, López (2008) proposed that during the late Holocene in the puna of Argentina the increase of demography and the spatial heterogeneity was followed by the decrease of the size of hunting social groups which also managed camelids in captivity. Based on the optimal group size model, individuals forming smaller groups maximized their return rates along more restricted habitats, setting the stage for the development of pastoral economies. Under this same approach, prestige- technologies and rock art scenes displaying social distinctiveness were interpreted as the archaeological evidence of the emergence of social orders based on social heterogeneity and leaderships associated with the first food producers societies in the northern puna of Argentina (Muscio 2006). In this model, social heterogeneity and leadership resulted from intragroup competition. Importantly, the shift towards smaller social groups discussed by López, and the social behavior of recognizing leaderships for ordering intragroup competition discussed by Muscio, both are part of the behavioral plasticity of humans regarding social behavior and decision making, created and maintained by selection. Then, if we grant that the Human Evolutionary Ecology provides tools for studying the emergence and the adaptive design of technologies and complex behavioral patterns at behavioral time scales (Smith 2001), we must also give a causal role to the selective forces of cultural transmission for the preservation of this adaptive variation at the scale of the population and in the long run. Because Human Evolutionary Ecology is silent regarding how adaptive variation is preserved along evolutionary time frameworks, we must introduce the action of evolutionary mechanisms as transmission and selection for a complete explanation of this critical issue.

For instance, adaptive technologies are expected to follow predictable patterns of convergence as environmental conditions are similar. In this way we can appeal to models based on the Human Behavioral Ecology approach for studying these convergences in adaptive design. Technologies that have a high potential for convergence were studied in Patagonia by Cardillo et al (2013) where it is documented that hunter-gatherer groups which were distanced for more than 1000 km shared similar technological adaptations regarding the harvesting of marine resources. On the basis of our current knowledge about human adaptive learning and its effects on cumulative cultural evolution, we must conclude that the long-term *persistence* of these technological convergences at the scale of the population is mediated by the action of cultural transmission, and not by the continual reinvention of this variation as a pure decision making process would imply.

In this way, *invention* (a novel goal-oriented produced cultural variation) and *innovation* (a retained invention) are useful heuristic concepts in cultural evolution. For instance, technological innovations have been documented by López (2009), and López and Restifo (2009), in the northern Puna of Argentina. In this case, processes of adaptive cultural innovation have been invoked to explain the emergence and spread of laminar lithic technology during the mid Holocene in a context of increased population density, economic intensification, and increased social groups. This phenomena is absent in the southern puna of Argentina, suggesting a process of adaptive cultural divergence between the two regions. Moreover, the increasing diversification of lithic artifacts classes during the mid Holocene in the puna region was explained as an effect of an accelerated rate of cultural evolution dependent on an increasing demography, compared with the low rate of cultural innovations recorded during the early Holocene in small and fluctuating population contexts Muscio (2011)

In short, culture contains all the elements of a Darwinian evolutionary system, variation, inheritance, and differential replicative success; which results a) from the impact of the cultural variation on the probabilities of the individuals for transmitting their cultural variants to others; b) from the existence of replicative biases favoring some variants over others (cultural selection), and c) by chance events that interfere cultural transmission producing neutral evolution. Moreover, humans are a proved behavioral flexible species, from which a diversity of traits and behaviors as adaptive responses to varying ecological contexts results. Based on all these elements, Evolutionary Archeology has a broad and ambitious agenda: the explanation of the archaeological record in terms of the Darwinian theory of evolution by clarifying the causes and consequences of the evolutionary processes.

By its emphasis on blind variation and Mendelian inheritance, Neo-Darwinism is not an accurate framework for theory building on cultural evolution. Instead, considering the particularities of cultural inheritance, Darwinism with its focus on the differential persistence of variability is the proper theoretical ground for the study of emergence and evolution of cultural variation at broad spatial and temporal scales. The construction of a synthetic paradigm in archaeology based on the Darwinian model of evolution does not demand the suppression of theoretical diversity (Muscio 2009). Conversely, the consolidation of a common theoretical ground as the basis for the construction of a variability of models and alternative hypotheses which ultimately will fuel the natural selection of competing scientific ideas is required. This task needs a robust theoretical structure with middle range statements, that enables us to link the properties of the archaeological record with the complexities of the Darwinian theory of evolution.

About this book

Following Gould (2002) convergence, divergence, and parallelism are of critical importance in any evolutionary research agenda. From this point of view, the *introduction and persistence* of cultural variation along large time scales becomes archaeology a unique field for the study of human evolution by selection and other evolutionary processes by which convergence, divergence or parallelism arise in cultural lineages. As we have showed, the study of evolution in the archaeological record is based on the empirical fact that culture is an inheritance system, which is independent (but also related) from the genetic inheritance system. As such, cultural inheritance evolves forming cultural lineages.

The work of Santiago Ginnóbili, from an epistemological point of view discusses the nature of variation and its relation to the mechanism of natural selection described by Darwin itself, not by the Neo Darwinism of the modern synthesis. In particular, the paper focuses on the logical status of the directed variation by assessing the ways by which natural selection acts upon it. The work is an important contribution for those interested in cultural evolution, as it illuminates the fact that even when cultural evolution has Lamarckian instances this does not imply that it cannot evolve by natural selection.

As variation is central to evolution, the documentation of *variation* along space and time is a way for studying evolutionary processes in archaeology. Most of the papers presented in this volume address this issue. The building of cultural phylogenies is an important way to approach this question. Of course, the documentation of culture phylogeny raises new methodological challenges related to the properties of cultural inheritance as well as to the techniques used in phylogenetic reconstruction. Many of these challenges are creatively addressed by the authors in this book.

In his work, Daniel García Rivero critically introduces the main classification methods that are useful in Evolutionary Archaeology for phylogenetic reconstruction. The paper is concerned with the discussion of the assumptions of different philosophical schools of taxonomy: phenetics, cladistics, and evolutionary taxonomy, concluding that there is not consensus to prefer some method over another in Evolutionary Archaeology. In this vein the paper of Marcelo Cardillo also discusses phylogenetic reconstruction methods but on empirical grounds. The author also appeals to multivariate methods for detecting patterning in the distribution of artifactual variation by using parsimony as the epistemological basis to build cultural phylogenies through cladistics. This method is used in order to explore the evolutionary role of three dimensions: space, environment, and history, as sources of causality for explaining the technological diversity of the Patagonian coast.

Cladistics is a method for *hypotheses building* regarding evolutionary relatedness among taxa, and it is based on the principle of maximum parsimony- the preference of the simpler of two or more otherwise equally adequate explanations. In cultural evolution relatedness rests on cultural transmission. Hence, the first critical task in cultural phylogenetic reconstruction is the recognition of homologies: traits which are similar due to descent from a common ancestor. Cladistics reconstructs through dichotomous diagrams the successive changes of a set of traits grouped in classes which are more or less evolutionary related by common descent (see Farris 1983, 1986, Forey et al. 1992). Thus, cladistics traces evolution at the scale of the traits. Because by cladistics the separation of clades (monophyletic groups) is modeled as a branching process, the method follows the Darwinian pattern of divergent evolution consisting in descent with modification. Cladistics differs from other methods such as phenetic, because classes (terminal taxa) are grouped together based on whether or not they have one or more shared unique trait that come from the group's last common ancestor and are not present in more distant ancestors (*synapomorphies)*.

As García Rivero notes in this book, this is not the case of phenetics which focuses on overall similarity (see discussion in Sneath y Sokal 1973). However, as discussed by Rivero (in this volume) and Cardillo (in this volume), both methods, cladistics and phenetics, can be used comparatively in order to build phylogenetic hypotheses in cultural evolution under different assumptions (see also Shennan and Collard 2005). As was stated by Makarenkov and Legendre (2004), and by Sneath y Sokal (1973) at an intraclass level (for example, when comparing the variation within a class) we may be more interested in tracing the gradients of variation along a phylogeny than in the qualitative changes of traits. In this case, phenetic techniques (see Rivero, this volume) were developed for the quantitative study of evolution and are widespread, for example, in ecology and genetics.

But not only statistical appropriate methods to model character evolution are important, also the selection of correct cultural traits is critical. The paper of Gustavo Barrientos addresses the difficulties associated with the recognition of homologous traits in cultural datasets and its derivations for studying cultural phylogenies. After an insightful discussion of the potential sources of error for the identification of homologies, Barrientos calls for the building of a theoretical framework for homology recognition and testing by adopting a developmental perspective, based on the study of ontogenetic trajectories of artifacts on an individual and population basis. In this way the work of Barrientos contributes to the methodological side of phylogenetic reconstruction and its potentials for inferring macroevolutionary patterns of cultural change.

At the macroevolutionary scale, cultural evolution is traceable by the patterns of divergence, parallelism, or convergence documented by phylogenetic reconstructions and other methods (Gould 2002, see Collard and Shennan 2006, Cardillo and Charlin 2013). Shortly, divergent evolution is the accumulation of differences among human populations or artifacts classes which lead to the formation of new derived classes. Parallel evolution is the acquisition of a similar trait in related, but distinct, human populations or artifact classes descending from the same ancestor, but from different clades. Convergent evolution is the acquisition of the same trait in unrelated cultural lineages. For instance, a gradual process of cultural divergence and adaptive radiation has been proposed by Borrero (1989-1990) in Tierra del Fuego. Also Charlin et al. (2013), by using statistical techniques to the comparative study of the variability in the archaeological record, showed evidence of a process of cultural divergence between Tierra del Fuego and continental southern Patagonia. Convergent behavioral patterns often result from purely adaptive processes by which human groups respond in predictable ways confronting similar environmental pressures. Adaptive plasticity produces these phenomena in human behavior. Hence, Human Behavioral Ecology is a well informed theoretical framework for studying convergent evolution when the processes explaining the long term persistence of the environmentally induced convergences are considered.

In phylogenetic reconstruction, convergent evolution is a source of homoplasy, as well as parallelism and character reversals to an ancestral form. If the amount of homoplasy is high for a given dataset we say that the cultural dataset lacks a phylogenetic signal, given the absence of a cladistic (branching) structure in the dataset. Broadly, homoplasy is a character that is shared by multiple taxa due to some cause other than common ancestry (see Barrientos this volume). Hence, homoplasy documents other processes but not decent with modification. Another important source of homoplasy is horizontal cultural transfer. This happens when unrelated cultural lineages exchanges traits. For example, in Northwestern Argentina, during the Inca expansion, local ceramic traditions adopted traits from the Inca ceramic traditions. The results of these processes are reticulated phylogenies, such as those characterizing many plants and bacterial evolutionary history, where gene horizontal transfer is common. In these cases, the analytical research may consist in the documentation of the background pattern of decent with modification by removing these homoplasious traits from the dataset, and then to proceed exploring the reticulated phylogenies in order to gain information about the historical processes that produced them (Muscio 2011). At any rate, phylogenetic reconstruction should start with independent information about the homologous status of the variation.

Interestingly, as Cardillo and Charlin (2010, 2013) show, horizontal cultural transfer is a powerful way to reduce artifactual divergence. As the authors documented on broad spatial scales, the correlation between spatial and phylogenetic information is informative of gradients of morphological change in the projectile points of the mid-late Holocene. This pattern is causally linked to isolation by distance and to processes of information flow among populations. Both processes resulted in a low morphological divergence and parallelism in the evolutionary pathway of the projectile points of northern and southern Patagonia. On this basis the study of reticulated phylogenies is a necessary step towards illuminating the broad spatial processes of transmission that eroded cultural divergence. In addition, the patterns of spatial and temporal distribution of cultural lineages can document processes like population expansions or population extinctions. The phylogenetic analysis of datasets coming from large spaces can be useful to document cladistic patterns correlated with geography.

In this vein, the paper of Hernán Muscio, from a perspective based on population dynamics models the role of localized extinction in the cumulative evolution of cultural variation among subpopulations connected by migration. On this basis, it is suggested that group cultural differences may be lost as a consequence of a high rate of local extinction and migration. The analytical results are used to predict the archaeological patterns of representation of cultural lineages through time and space in ecological settings where localized extinction is frequent. By using cladistics, the author suggests that local extinction played an important role in the spread and evolution of cultural lineages among the earliest populations with economic niches based on high altitude agriculture in the northern puna of Argentina.

The processes of local extinction have received little attention by evolutionary archaeologists. Localized extinction is expected particularly during the colonization of new environments when individuals face unfamiliar environments. From a demographic point of view, smaller populations are more prone to local extinction by demographic stochasticity. From an ecological point of view, local extinction will be common in fluctuating environments and low quality habitats. An important question is the way in which local extinction occurs as the result of selection. This is called Darwinian extinction and takes place when a population under natural selection evolves along a path to extinction. For example, if optimizing selection leads to a increasing consumption of a small number of preys, a selective feedback between hunting behavior and a diminishing prey population can lead to the local extinction of the human population by the overhunting of its resource base. Hence, when we leave aside the adaptationist idea that phenotypic plasticity immunizes humans from the action of selection, we can study potential sources of past extinctions and potential mismatches between populations and their environments.

The paper by Diego Rivero and Matías Medina, based on the Human Behavioral Ecology approach, focuses on the consequences of the human colonization of the Sierras of Cordoba (Argentina) on the populations of high-ranked resources impacted by hunting pressures. The authors explore the long-term consequences of human induced reductions of high quality resource stocks for the evolution of broad-spectrum economic adaptations. Importantly, the authors conclude that a broad-spectrum diet may not have been an exclusively Late Holocene phenomenon and its origins may be found in earlier processes beginning during the Middle Holocene. Thus, the incorporation of small-scale agriculture around ca. 1200-1000 BP cannot be satisfactorily explained only by external factors such as climate changes or the diffusion and/or arrival of new biological populations. On this basis, Rivero and Medina call to consider the agency of past human cultural activities in process of niche construction, co-directing subsequent cultural changes that are archaeologically documented. Among the merits of this work it is the use of the rationale of the optimization models for illuminating long-term evolutionary processes, like niche construction.

In the same vein, Belardi et al. explore from a biogeographical framework the integration of several Darwinian approaches as we have shown previously. The authors focus on models of cultural transmission and particularly on niche construction (Odling-Smee 2003). On this basis cultural transmission is conceived as a key element in niche construction in large temporal and spatial scales. Based on the distributional patterns of rock art motif and considering its properties at broader spatial scales, the paper models important processes such as demographic evolution, mobility, and contact between groups. In this line, the authors show how the way in which human populations in Patagonia modified their environments by using rock art for cultural transmission and space signalling. In this way, the paper demonstrates the benefits of using different lines of evidence in order to discuss past evolutionary processes in archaeology.

Finally, as this book shows, Evolutionary Archaeology in Argentina is a consolidated field of research which during the last years has grown by the sophistication of its theoretical basis and the development of new methodologies for studying evolution in the archaeological record.We celebrate the way this intellectual tradition currently evolves in the Southern portion of América; the land that 150 years ago fuelled Darwin´s brilliant inspiration.

Aknowlegements: To Karen Borrazo for commentaries and sugerencies of the earliest version

References

Barberena, R. 2002. Los límites del mar. Isótopos estables en Patagonia meridional. Sociedad Argentina de Antropología. Buenos Aires. Argentina.

-2005. Fronteras en tiempo arqueológico. In: *En La frontera. Realidades y representaciones.* CONICET-IMHICIHU, pp. 33-46. Buenos Aires.

-2008. Arqueología y biogeografía humana en Patagonia Meridional. *Sociedad Argentina de Antropología.* Buenos Aires.

Binford, L. 1962. Archaeology as Anthropology. *American Antiquity*, 28: 217-225

Borgerhoff Mulder, M., Nunn, C.L., Towner, M.C., 2006. Cultural macroevolution and the transmission of traits. *Evolutionary Anthropology* 15, 52-64.

Borrero, L.A. 1989. Replanteo de la arqueología patagónica. *Interciencia* 14 (3): 127-135.

-1989-1990. Evolución cultural divergente en la Patagonia Austral. *Anales del Instituto de la Patagonia*, 19:133-40.

-1993. Site Formation Processes in Patagonia: Depositional Rates and the Properties of the Archaeological Record. In: *Explotación de recursos faunísticos en sistemas adaptativos americanos.* L.A. Borrero and J.L. Lanata, (eds.). Arqueología Contemporánea 4: 107-122.

Borrero, L. A. 1994-95. Arqueología de la Patagonia. Palimpsesto, Revista de Arqueología 4: 9-56.

-1992. Artefactos y evolución. *Palimpsesto. Revista de Arqueología* 3: 15-32.

-1944-1995. Arqueologia de la Patagonia. *Palimsesto. Revista de Arqueología* 4: 9-69.

Borrero, L.A., R. Barberena, N. V. Franco, J. Charlin, R.H Tykot. 2009. Isotopes and Rocks: Geographical Organisation of Southern Patagonian Hunter-Gatherers. *International Journal of Ostearchaeology* 19(2). In press.

Boschin, M. T. and A. M. LLamazares.1986. La escuela histórico-cultural como factor retardatario del desarrollo científico de la arqueología argentina. *Etnía* 32.

Boyd, R. and P.Richerson. 1985. *Culture and the Evolutionary Process.* Chicago, University of Chicago Press.

-2005. Not By Genes Alone: *How Culture Transformed Human Evolution.* Chicago, University of Chicago Press.

Camino, U. 2009. La Cerámica Del Período Agro-Alfarero Temprano Como Estrategia Evolutiva en La Quebrada de Matancillas (Puna De La Provincia De Salta). In: Arqueología Y Evolución: Teoría, Metodología y Casos de Estudio, G. López y M. Cardillo (eds.), pp. 161-197

Cardillo, M. 2002. Transmisión Cultural y Persistencia Diferencial de Rasgos. Un Modelo para el Estudio de la Variación Morfológica de las Puntas de Proyectil Lanceoladas de San Antonio de los Cobres, Provincia de Salta, Argentina. In: *Perspectivas Integradoras entre Arqueología y Evolución. Teoría Método y Casos de Aplicación*. Martínez GA and Lanata JL (eds.) INCUAPA. Serie teórica 1: 97-199.

Cardillo, M. 2009. Temporal Trends in the Morphometric Variation of the Lithic projectile Points during the Middle Holocene of Southern Andes (Puna Region). In: Muscio, H. J. y G. López (eds.), *Theoretical and Methodological issues in Evolutionary Archaeology. Toward a Unified Darwinian Paradigm*. Oxford, British Archaeological Reports International Series 1915, pp 13-20.

Cardillo, M and J. Charlin. 2010. Diversificación morfológica en las puntas de proyectil de Patagonia. Explorando dimensiones espaciales y temporales. *Primer Encuentro de Morfometría "Morfometría Geométrica y Estudios Filogenético*. La Plata. Buenos Aires. MS.

Cardillo M and J. Charlin. 2013. Morphological diversification of stemmed projectile points of Patagonia. Assesing spatial and environmental patters by means of phylogenies and comparative methods. *Multidisciplinary Approaches to the Study of Stone Age Weaponry*. Springer. In Press.

Cardillo, M.; F. Scartascini and P.A. Zangrando.2013. Diferencias en el diseño y funcionalidad de las pesas líticas de la costa norte de Patagonia y el canal de Beagle, Tierra del Fuego. Un enfoque morfométrico y cuantitativo. MS.

Charlin J, K. Borrazo, M.Cardillo. 2013. Exploring size and shape variations in late Holocene projectile points in Northern and Southern coasts of Magellan Strait (South America). *British Archaeological Reports*. In Press.

Collard, M., Stephen, T., Shennan, Jamshid, J., Tehrani, J., 2006. Branching, blending, and the evolution of cultural similarities and differences among human populations. *Evolution and Human Behavior* 27: 169-184

Darwin C. 1871. *The descent of man, and selection in relation to sex*. 7th thousand. London: John Murray. Vol. 1.

Dunnell, R. C. 1978. Style and Function: A fundamental dichotomy. *American Antiquity* 43:192-202.

-1980. Evolutionary Teory and Archaeology. *Advances in Archaeological and Theory* 3:38-100.

-1989. Aspects of the application of evolutionary theory in archaeology. In *Archaeological Thought in America*. C. Lamberg-Karlovsky (eds.) Cambridge, Cambridge University Press, pp. 35-49.

-2006 Measuring Relatedness. In: *Mapping Our Ancestors: Phylogenetic Approaches in Anthropology and Prehistory*, C.P. Lipo; M. J. O'Brien, M. Collard, S. J. Shennan (eds.), pp. 109-118. Aldine, New York.

Durham, W. 1992. *Coevolution: Genes, Culture and Human Diversity*. Stanford, Stanford University Press.

Farris, J. S. 1983. The logical basis of phylogenetic analysis. In: Platnick, N. I. y V. Funk (eds.), *Advances in Cladistics* 2, New York Botanical Garden, New York, pp. 7-36.

-1986. Synapomorphy, parsimony, and evi-dence. *Taxon* 35: 298-315

Forey, P. L.; C. J. Humphries.; I. J. Kitching.; R. W. Scot-land.; D. J. Siebert and D. M. Williams (eds.) 1992. *Cladistics: A practical course in Systematics*. Clarendon Press, Oxford Science Publications, The Systematics Association Publication no. 10, Oxford.

Ghalambor, C.K.; J.K. McKay.; S.P. Carroll and D.N. Reznick. 2007. Adaptive versus non-adaptive phenotypic plasticity and the potential for contemporary adaptation in new environments. *Functional Ecology* Vol 21(3):394-407.

González, A. R. 1963. Las Tradiciones Alfareras del Período Temprano del N.O Argentino y Sus Relaciones con las áreas aledañas. *Congreso Internacional de Arqueología de San Pedro de Atacama*. Anales De La Universidad de Chile, 2. Antofagasta.

-1979. Dinámica Cultural Del N.O Argentino Evolución e Historia En Las Culturas Del N.O. Argentino. *Antiquitas*. 28-29: pp 1-15

Gould, S. J. 1994. Tempo and Mode in the Macroevolutionary Reconstruction of Darwinism. *Proceedings of the National Academic of Sciences of the Unites States of America* 91: 6764-6771.

-2002. *The Structure of Evolutionary Theory*. Washington, The Belknap Press of Harvard University Press.

Haldane, J.B.S. 1957. The Cost of Natural Selection. *J. Genet.* 55: 511-524.
Henrich, J. 2004. Demography and Cultural Evolution: Why adaptive cultural processes produced maladaptive losses in Tasmania. *American Antiquity*, 69 (2): 197-21
Hughes, S. L. 1996. Getting to the Point: Evolutionary Change in Prehistoric Weapondry. *Journal of Archaeological Method and Theory* 5 (4):345-408.
Kronfeldner, M. E. 2007. Is cultural evolution Lamarckian? *Biology & Philosophy*, Volume 22 (4): 493-512
Lanata, J. L. 1995. *Paisajes Arqueológicos y Propiedades del Registro en el Sudeste Fueguino*. Tesis Doctoral. Facultad de Filosofía y Letras. UBA.MS.
-1996. Diversidad Instrumental en el Norte de Península Mitre, Tierra del Fuego. *Arqueología* 6, pp. 159-197.
Lanata, J. L. y Borrero; L. A. 1994. Riesgo y Arqueología. *Arqueología de Cazadores Recolectores. Límites Casos y Aperturas*. J. L. Lanata and L. A Borrero compilators. Arqueología Contemporánea 5: 129-143
Lewontin, R.C. 1983. The organism as the subject and object of evolution, repr. In: *The Dialectical Biologist*. R. Levins and R. C. Lewontin (eds.)Harvard University Press, Cambridge, MA, pp. 85-106.
López, G. 2009. The Study of the Archaeological record of Santa Rosa de los Pastos Grandes, Puna of Salta, Argentina, from an Inclusive Evolutionary Perspective. In *Theoretical and Methodological issues in Evolutionary Archaeology*. Muscio, H. J. y G. López (eds.), Toward a Unified Darwinian Paradigm, pp. 49-55. Oxford, British Archaeological Reports International Series 1915.
López, G.; F. Restifo. 2009. Modelando el cambio en la toma de decisión tecnológica desde una perspectiva evolutiva: expectativas arqueológicas para el análisis en Pastos Grandes, Puna de Salta. In: López, G., Cardillo, M. (eds.), Arqueología y Evolución. Teoría, metodología y casos de estudio. SB, Buenos Aires, pp. 109-140.
Loponte, D.; A. Acosta and J. Musali. 2006 Complexity among Hunter-Gatherers from Pampean Region, South America. In: *Beyond Affluent Forager: Rethinking Hunter-Gatherers Complexity*, Colin Grier, Jangsuk Kim and Junzo Uchiyama (eds.) pp. 106–125. Oxbow Books, Oxford.
Makarenkov, V., Legendre, P., 2004. From a Phylogenetic tree to a reticulated network. Journal of Computational Biology 11: 195-212.
Medina. M. 2008. *Diversificación Económica y Uso del Espacio en el Tardío Prehispánico del Norte del Valle de Punilla, Pampa de Olaen y Llanura Noroccidental (Córdoba, Argentina)*. Tesis Doctoral. Facultad de Filosofía y Letras, Universidad de Buenos Aires, Buenos Aires. MS.
Muscio, H.J. 1999. Colonización humana del NOA y variación en el consumo de recursos: la ecología de los cazadores- recolectores de la puna durante la transición Pleistoceno-Holoceno. *http//www.naya.org.ar/art%23237%3Bculos/index.htm.*
-2004. *Dinámica Poblacional y Evolución Durante el Período Agroalfarero Temprano en el Valle de San Antonio de los Cobres, Puna de Salta, Argentina*. Tesis Doctoral, Facultad de Filosofía y Letras, Universidad de Buenos Aires. MS.
-2006. Una aproximación evolutiva a la complejidad y al orden social durante el período temprano a través del estudio de representaciones rupestres de la quebrada de Matancillas (Puna argentina). *Estudios Atacameños* 31: 9-30.
-2011. Modelling demographic dynamics and cultural evolution: The case of the early and mid-Holocene archaeology in the highlands of South America. *Quaternary International* 30:1-8.
Neme G. 2009. Un enfoque regional en cazadores-recolectores del oeste argentino: el potencial de la ecología humana. In *Perspectivas Actuales en Arqueología Argentina*. CONICET – IMHICIHU.

Odling-Smee, F. J., K. N. Laland and M. W. Feldman. 2003. Niche Construction. The Neglected Process in Evolution. *Monographs in Population Biology* 37. Princeton, Princeton University Press.
Olivera, D. E. 1988. La Opción Productiva: Apuntes para el análisis de sistemas adaptativos de tipo formativo del Noroeste Argentino. In: *Precirculados de las ponencias científicas a los simposios del IX Cong. Nac. De Arqueología Argentina* , pp. 83-101. ICA Buenos Aires
Ratto, N. 1994. Funcionalidad vs. Adscripción cultural: cabezales líticos de la margen norte del estrecho de magallanes. *Arqueología de cazadores-recolectores. Límites, casos y aperturas*. J.L. Lanata y L.A Borrero compilators. Arqueología Contemporánea 5:105-120

Rindos, D. 1989. Diversity, variation and selection. En Quantifying Diversity in Archaeology. R.D Leonard y G.T Jones (eds). New Directions in Archaeology. Cambridge University Press. Cambridge.

Scartascini, F. and M. Cardillo. 2009. Explorando la variabilidad métrica y morfológica de las "pesas líticas" recuperadas en el sector norte de la costa del golfo de San Matías. *Arqueometría Latinoamericana : Segundo Congreso Argentino y Primero Latinoamericano*. Vol 1: 162-168.

Rogers, D.S.; P.R. Ehrlich. 2008. Natural selection and cultural rates of change. Proceedings of the National Academy of Sciences 105: 3416-3420.

Scheinsohn, V. 1997. *Explotación de materias primas óseas en la Isla Grande de Tierra del Fuego*. Unpublished PhD Thesis, University of Buenos Aires.

-2002. Un modelo evolutivo en Argentina. Resultados y perspectivas futuras. In: G. Martínez and J.L. Lanata (eds.), *Perspectivas Integradoras entre Arqueología y Evolución. Teoría, Método y Casos de Aplicación*. Olavarría, Universidad Nacional del Centro de la Prov. de Bs. As. (UNCPBA), INCUAPA, Serie Teórica N°1, 187-206.

-2008. Andando el Carro se acomodan los zapallos. Puentes hacia el pasado, reflexiones teóricas en arqueología. D. S Jackson, D Salazar, A Troncoso (eds.). *Monografica de la Sociedad Chilena de Arqueología.*

- 2010. The Good, the Bad and the Ugly: Harpoon Heads of Beagle Channel, Isla Grande de Tierra del Fuego (Patagonia, Argentina). In: *Ancient and Modern Bone Artefacts from America to Russia. Cultural, Technological and Functional Signature*, (eds.) A. Legrand-Pineau.; I. Sidera, N. Buc.; E. David and V. Scheinsohn, pp. 295-302. Oxford, Bar International Series.

Shennan, S. J. 2002. *Genes, Memes and Human History: Darwinian Archaeology and Cultural Evolution*. London: Thames and Hudson.

Shennan, S.J. and M. Collard. 2005. Investigating Processes of Cultural Evolution on the North Coast of New Guinea with Multivariate and Cladistic Analyses. Chapter 8. In: *The Evolution of Cultural Diversity: A Phylogenetic Approach* R. Mace; C.J. Holden and S.J. Shennan (eds.) London: UCL Press, 133-164

Smith. E. A.; M. Borgerhoff-Mulder and K. Hill. 2001. Controversies in the evolutionary social sciences: A guide to the perplexed. *Trends in Ecology & Evolution* 16:128-135.

Sneath P.H.E. and Sokal R.R. 1973. *Principles of Numerical Taxonomy*. Freeman & Co., San Francisco

Thomas, H. H. 1989. Diversity in hunter-gatherer cultural geography. In: *Quantifyng Diversity of Archaeology*. R.D. Leonard y G.T. Jones (eds.) Cambridge University Press. Cambridge, 85-91.

West-Eberhard, M.J. 2003. *Developmental Plasticity and Evolution.* Oxford University Press, New York.

Yacobaccio, H. D.; C. Madero and M. Malmierca. 1998. Etnoarqueología de Pastores Surandinos. G.Z.C. Buenos Aires.

CULTURAL ADAPTATIONS: IS IT CONCEPTUALLY COHERENT TO APPLY NATURAL SELECTION TO CULTURAL EVOLUTION?

Santiago Ginnobili

Universidad de Buenos Aires – Universidad de Quilmes
e-mail: santi75@gmail.com

Abstract

Critics of Darwinian approaches to the study of the evolution of human and social behavior often express their dissent claiming that cultural evolution is Lamarckian. By this they mean two things. First, that contrary to natural selection, in Lamarckian mechanisms of use and disuse plus the inheritance of acquired traits, the modifications in an organism arise as a solution to the environmental problem at hand, i.e., variation is not blind; and second, that the acquired trait is transmitted to the offspring by imitation or learning.

In this paper I will reconstruct informally the theory of natural selection as it was used by Darwin in order to show that the fact that cultural evolution has these Lamarckian features does not imply that it cannot evolve by natural selection. The appeal to Darwin's original writings has two advantages. On the one hand, the way he applies the theory makes it more transparent than in contemporary applications which are its fundamental concepts and structure. On the other, Darwin holds that Lamarckian mechanisms can be causally responsible for variations on which natural selection operates, thus showing that it is possible to hold a theory of natural selection that is not incompatible with these alleged Lamarckian features of cultural evolution.

KEYWORDS: NATURAL SELECTION, CULTURAL EVOLUTION, DARWIN, STRUCTURE OF THE THEORY OF NATURAL SELECTION

Resumen

Los críticos de los enfoques que intentan estudiar la evolución conductual humana y social con enfoques darwinianos suelen describir tal rechazo con la afirmación: "La evolución cultural es lamarckiana". Con esto se refieren, por un lado, al hecho de que, supuestamente a diferencia de la selección natural, en los mecanismos lamarckianos de uso y desuso más la herencia de caracteres adquiridos, la modificación en el rasgo del organismo en cuestión surge como solución al problema ambiental en juego, la variación no es ciega, y por el otro, que el rasgo adquirido se transmite a la descendencia. Así, los rasgos culturales se proponen para solucionar ciertos problemas ambientales y son transmitidos a la descendencia por imitación o aprendizaje.

En este trabajo se reconstruirá informalmente la teoría de la selección natural tal como era utilizada por Darwin con el objetivo de mostrar que el hecho de que la evolución cultural tenga estas características lamarckianas no implica que no puede evolucionar por selección natural. La apelación a los escritos originales de Darwin tiene dos beneficios. Por un lado la forma en que aplica la teoría resulta más transparente que en las aplicaciones actuales cuáles son los conceptos fundamentales de la teoría y cómo es su estructura. Por el otro, Darwin sostiene que los mecanismos lamarckianos pueden ser causantes de la variación sobre la cual la selección natural funciona, mostrando que es posible sostener una teoría de la selección natural que no sea incompatible con estos supuestos rasgos lamarckianos de la evolución cultural.

PALABRAS CLAVE: SELECCIÓN NATURAL, EVOLUCIÓN CULTURAL, DARWIN, ESTRUCTURA DE LA TEORÍA DE LA SELECCIÓN NATURAL.

Introduction

There are different ways of applying the theories of Darwinian evolutionary biology to cultural evolution, that is, to evolutionary processes in which inheritance does not proceed through genetic information. There are, consequently, a number of critics of these attempts. Independently of the relevance of the reasons such critics have for rejecting the application of evolutionary theories to the cultural realm, they usually use the claim *that cultural evolution is Lamarckian* as their banner. The target of this attack is in particular the application of natural selection to the evolution of cultural traits.

This paper proceeds at a metatheoretical level of conceptual elucidation. Independently of how cultural evolution does in fact occur, which is of the interest of scientists in the corresponding areas, my intention is to analyze conceptually whether there is an inconsistency between some versions of Lamarckism and Darwinism. My aim is to show that, even if cultural evolution had certain Lamarckian features – in particular, if variations were directed or if the direct instruction of the environment was possible in a sense to be explained later – this would not imply that the theory of natural selection cannot be applied to it. Someone might consider, however, that if the variation over which natural selection is to operate is not blind, such operation is superfluous. I believe this happens because the theory has not been adequately presented. I will, therefore, put forward an informal reconstruction of the theory of natural selection as it was used by Darwin. I will show how this theory did not in any sense presuppose that variation was blind. In fact, for Darwin, Lamarckian mechanisms of use and disuse were some of the mechanisms responsible for the variation required for natural selection to work.

This work, therefore, falls within the scope of the reconstructive and elucidatory goals established by logical positivists (Hahn et al. 1929). The reconstructive tools they provided, however, have over the years turned out unsatisfactory. I will utilize the more powerful and sophisticated tools provided by metatheoretical structuralism (but in an informal fashion) (Balzer et al. 1987).

Lamarkian cultural evolution

Although he was not the first to hold this idea (e.g. Medawar 1960), Stephen Jay Gould has been one of the authors that have insisted the most on the Lamarckian character of cultural evolution:

> "Cultural evolution has progressed at rates that Darwinian processes cannot begin to approach. Darwinian evolution continues in Homo sapiens, but at rates so slow that it no longer has much impact on our history. This crux in the earth's history has been reached because Lamarckian processes have finally been unleashed upon it. Human cultural evolution, in strong opposition to our biological history, is Lamarckian in character. What we learn in one generation, we transmit directly by teaching and writing. Acquired characters are inherited in technology and culture " (Gould 1980: 83-84).

Or, in a more recent text:

> "Human culture has introduced a new style of change to our planet, a form that Lamarck mistakenly advocated for biological evolution, but that does truly regulate cultural change–inheritance of acquired characters. Whatever we devise or improve in our lives, we pass directly to our offspring as machines and written instructions. Each generation can add, ameliorate, and pass on, thus imparting a progressive character to our technological artifacts" (Gould 1993: 215-216).

These texts seem to point to the fact that cultural traits are acquired during ontogenetic processes of development and are transmitted to offspring nor through genetic material, but through learning. But if the claim that cultural evolution is Lamarckian only intends to state this, then it becomes trivial. Given that cultural evolution is defined as the evolution of traits that does not involve any kind of genetic change, the prase "cultural evolution is Lamarckian" means the same as "cultural evolution is cultural", and it is safe to assume that this is not the point (Kronfeldner 2007: 502). The main point has to do with the way in which the traits that are transmitted to the offspring are acquired. In the phrase "acquired traits are inherited" a more restricted sense of "acquired" is usually presupposed, one that expresses the acquisition of traits adequate to solve certain environmental problem as a response to the environmental problem itself (Kronfeldner 2007: 495). This is the reason why, for instance, Boyd and Richardson characterize the cultural process of problem solving as Lamarckian (Boyd et al. 1985, p. 82). Variation would not be blind, but would rather be guided by the resolution of the problem at hand (Fog 1999: 65-67). And this is the central point according to Gould himself:

> "Why can't organisms figure out what would do them good, develop those adaptative features by dint of effort during their lifetimes, and the pass those improvements to their offspring in the form of altered heredity? We call such a putative mechanism "Lamarckism" or "the inheritance of acquired characters"" (Gould 1996: 221).

Lamarck explains the adaptation of organisms to their environment mainly through two laws (Lamarck 1809, chap. VII):

First law: During the life of animals, they exercise the use of certain organs and others get disused. The used ones get strengthened and developed, the others get weakened. (Commonly called the "law of use and disuse of organs").

Second law: The small and gradual changes individuals of a species experience throughout their lives are transmitted to their offspring. (The "law of inheritance of acquired traits" in the restricted sense).

The use of a given organ to solve some problem peculiar to the environment is what triggers the development of the organ improving previous solutions. Although the law of use and disuse can hardly be applied literally to cultural evolution, it is this feature of the way that variation arises that seems to be in play here. Indeed, according to many authors, it would be an essential feature of the theory of natural selection that the variation over which it operates is random, blind or decoupled (e.g. Fracchia et al. 2005:17; Gomila 2009: 340; Gould 1996: 221).

According to Campbell, for variations to be blind they must meet the following requirements (Campbell 1974, p. 421):

1. They must arise without a prior knowledge of which ones, if any, are going to be selected for.
2. They are independent of the environmental conditions at the moment of their occurrence.
3. The occurrence of each trial is not correlated with the solution, that is, specific correct trials are not more probable that the rest.
4. A variation succeeding an incorrect trial is not a correction of it.

It can easily be seen why mechanisms as use and disuse, which presuppose the inheritance of acquired traits, would not be blind. They would not satisfy the second requirement, since environmental conditions would be causally responsible for successful variation, and neither the third, because successful variations would be more probable than unsuccessful ones. If the variation postulated by the theory of natural selection is blind, then the theory of natural selection is inconsistent with Lamarck's first law. Of course, it is precisely this blindness that has been found attractive by some authors in order to think of cultural change. For instance, it is this feature that drives Popper into considering the change of scientific theories as analogous to biological evolution. Just as there is no direct instruction from the environment in biological evolution, there is no inductive process from experience to the discovery of new theories in the logic of scientific research (Popper 1974, pp. 34-41). Moreover, the fact that the theory of natural selection can operate over variations that are relatively spontaneous is what gives it much more explanatory scope than Lamarck's theory. Since Darwin accepted these two Lamarckian laws (unlike other aspects of his theory, as the tendency towards complexity that according to Lamarck drove evolution), but considered that the majority of the traits adequate for the environment remained unexplained:

> "Naturalists continually refer to external conditions, such as climate, food, &c., as the only possible cause of variation. In one very limited sense, as we shall hereafter see, this may be true; but it is preposterous to attribute to mere external conditions, the structure, for instance, of the woodpecker, with its feet, tail, beak, and tongue, so admirably adapted to catch insects under the bark of trees. In the case of the misseltoe, which draws its nourishment from certain trees, which has seeds that must be transported by certain birds, and which has flowers with separate sexes absolutely requiring the agency of certain insects to bring pollen from one flower to the other, it is equally preposterous to account for the structure of this parasite, with its relations to several distinct organic beings, by the effects of external conditions, or of habit, or of the volition of the plant itself" (Darwin 1859, p. 3).

The reference to explanation by means of habit or the will points to the fact that use and disuse presuppose the will or the habit of using or not using certain organ.

Part of the explanatory power of the theory, then, lies in that it is not necessary to accept any mechanism of direct instruction by the environment in order for adaptation to such environment to be increased. But, is it necessary that variation in all cases be independent in this sense from the environment? Variation can be blind, but must it be so? If that was the case, then, indeed, the theory of natural selection would be inconsistent with direct instruction by the environment and, if cultural traits arise as a direct response to environmental problems, then the theory of natural selection could not be coherently applied to cultural evolution.

It is customary to characterize the theory of natural selection using the expression "blind variation and selective retention". I believe there are two reasons why it is affirmed that variation must be blind. On the one hand, it is because ordinary applications of the theory involve genetically determined traits, and current molecular biology has shown (though previous authors such as Weissmann had already suggested it) that there is no system that allows modifications acquired during the process of ontogenetic development to be inscribed into the genome that will be passed on to the offspring. The other reason is that, since there is not any good reconstruction and presentation of the theory of natural selection, it is not clear why a case of selection over variations produced in a non blind manner would count as a case of natural selection rather than of the operation of Lamarckian laws. Thus, for instance, Richards describes Toulmin's appeal to a non-blind or, in the latter's own terminology, "coupled" theory of natural selection to account for the evolution of conceptual populations (Toulmin 1972, pp. 324-340), as an abandonment of natural selection altogether and an acceptance of a Lamarckian mechanism (Richards 1987, p. 578).

Even those authors that are not committed to blind variation and that purport to provide a version of the theory of natural selection general enough as to allow for its application to cultural phenomena present it poorly as consisting of the three principles of variation, reproduction and selection, without much elucidation of such concepts and the way they are applied (Álvarez 2009: 321; Fog 1999: 60). This makes room for unfounded suspicions about whether, in effect, it is the same theory which is operating. I will now present a reconstruction of the theory of natural selection that, though informal and sketchy, suffices for showing that this theory is more complex than the mere variation and selection. The presentation of its fundamental concepts, the form of its fundamental law and the different ways in which the latter is applied in its special laws, will erase any fear that accepting directed variations for natural selection destroys what is peculiar to the theory.

The Darwinian Theory of Natural Selection

As is well known, what Darwin wants to explain with the theory of natural selection (TNS) is certain adequacy or adjustment of organisms to their environment.

For instance:

> "The giraffe, by its lofty stature, much elongated neck, fore-legs, head and tongue, has its whole frame beautifully adapted for browsing on the higher branches of trees. It can thus

> obtain food beyond the reach of the other Ungulata or hoofed animals inhabiting the same country; and this must be a great advantage to it during dearths" (Darwin 1872, p. 177).

The way Darwin explains the fixation of this trait in the population of giraffes is the following:

> "So under nature with the nascent giraffe, *the individuals which were the highest browsers and were able during dearths to reach even an inch or two above the others*, will often have been preserved; for they will have roamed over the whole country in search of food. That the individuals of the same species often differ slightly in the relative lengths of all their parts may be seen in many works of natural history, in which careful measurements are given. These slight proportional differences, due to the laws of growth and variation, are not of the slightest use or importance to most species. But it will have been otherwise with he nascent giraffe, considering its probable habits of life; *for those individuals which had some one part or several parts of their bodies rather more elongated than usual, would generally have survived. These will have intercrossed and left offspring*, either inheriting the same bodily peculiarities, or with a tendency to vary again in the same manner; whilst the individuals, less favoured in the same respects, will have been the most liable to perish" (Darwin 1872, pp. 177-178, my emphasis).

If we consider, following metatheoretical structuralism, that TNS's fundamental law is that statement in which the fundamental concepts of TNS appear related (Balzer, Moulines & Sneed 1987, p. 19), we can extract an instantiation of the fundamental law of TNS from the previous passage:

Giraffes with the longer necks, fore-legs, heads and tongues are more effective in feeding from the higher branches of trees, thus improving their survival and, consequently, their differential reproductive success.

Nothing seems superfluous in this statement. If we removed any if its parts, the explanation it provides would falter. If the function assigned to the trait by the organism, that of reaching the higher branches of trees, is removed, we would not know why such trait could improve survival. It could improve it for other reasons, for instance, by intimidating possible predators. This would be an alternative and competing explanation to the one offered by Darwin. If we remove the improvement in survival, the relation between the trait and reproductive success would remain indeterminate. As we will see, this connection is not always established by improvement in survival.

We can find this very explanatory structure in other places in the *Origin*. For instance, it is possible to give a similar explanation in answer to the question "How was it that certain population of caterpillars acquired shapes similar to those of the branches in which they feed that allow them to pass unnoticed in order to protect them from predators?"

> "But in all the foregoing cases the insects in their original state no doubt presented some rude and accidental resemblance to an object commonly found in the stations frequented by them. Nor is this at all improbable, considering the almost infinite number of surrounding objects and the diversity in form and colour of the hosts of insects which exist. As some rude resemblance is necessary for the first start, we can understand how it is that the larger and higher animals do not (with the exception, as far as I know, of one fish) resemble for the sake of protection special objects, but only the surface which commonly surrounds them, and this chiefly in colour. Assuming that an insect originally happened to resemble in some degree a dead twig or a decayed leaf, and that it varied slightly in many ways, then all the variations which rendered the insect at all more like any such object, and thus favoured its escape, would be preserved, whilst other variations would be neglected and ultimately lost; or, if they rendered the insect at all less like the imitated object, they would be eliminated" (Darwin 1872, p. 182).

In this case the law-like statement presupposed is the following:

Caterpillars whose shape and color allow them to camouflage with the plant on which they feed tend to leave more descendants in virtue of having improved their survival in their environment.

Abstracting from these two statements we come closer to what I consider to be the fundamental law of TNS:

Individuals with traits that perform certain function more efficiently improve their survival, thus improving their differential reproductive success.

The fundamental law of TNS would have at least three components:

- The trait that performs certain function more efficiently.
- The differential reproductive success.
- The connection between the proper trait and reproductive success, which in these cases obtains by means of an improvement in survival.

To arrive at a more general version of the fundamental law of TNS we have to take into account that there are explanations that possess the same structure, but in which the connection between the proper trait and the improvement in differential reproductive success does not obtain by way of an improvement in survival. For instance, in the following case of sexual selection:

> "Generally, the most vigorous males, those which are best fitted for their places in nature, will leave most progeny. But in many cases, victory will depend not on general vigour, but on having special weapons, confined to the male sex. A hornless stag or spurless cock would have a poor chance of leaving offspring. Sexual selection by always allowing the victor to breed might surely give indomitable courage, length to the spur, and strength to the wing to strike in the spurred leg [...]" (Darwin 1859, p. 88).

The law-like statement presupposed in this case would be:

Roosters with spurs more effective for fighting against other roosters tend to mate more, consequently improving their differential reproductive success.

In other cases the explanation may resort neither to improvements in survival nor improvements in the ability to attract sexual partners. For example:

> "Those individual flowers which had the largest glands or nectaries, and which excreted most nectar, would be oftenest visited by insects, and would be oftenest crossed; and so in the long-run would gain the upper hand" (Darwin 1859, p. 92).

The law-like statement presupposed would be:

Plants that produce flowers more attractive to insects tend to increase their fertility improving, consequently, their success in differential reproduction.

The concept that varies through the different applications Darwin makes of his theory is proposed with TNS to explain what it purports to. It is an abstract concept that receives different interpretations and that affords Darwin many different explanations. The connection between the trait that is adequate for the environment and success in differential reproduction does not always obtain by means of an improvement in survival, as is generally supposed. If we call this concept "aptitude", the fundamental law of TNS could be:

Individuals with traits that perform certain function better improve their aptitude, thus improving their success in differential reproduction.

These are some of the fundamental concepts of the theory. There are others that I have not considered, such as that of environment or that of inheritance. But with this presentation it can already be seen that the structure is more complex than is usually acknowledged, as much in textbook presentations as in available reconstructions.

We can find different instantiations of this structure in the different special laws that arise from the several instantiations of the concept of *aptitude* throughout Darwins's writings. It is interesting to notice that this very same wide explanatory structure, that does not always include an improvement in survival, can already be found in Darwin's earliest evolutionist writings. For example:

> "[...] if the number of individuals of a species with plumed seeds could be increased by greater powers of dissemination within its own area (that is if the check to increase fell chiefly on the seeds), those seeds which were provided with ever so little more down, or with a plume placed so as to be slightly more acted on by the winds, would in the long run tend to be most disseminated; and hence a greater number of seeds thus formed would germinate, and would tend to produce plants inheriting this slightly better adapted down" (Darwin 1844: 92).

In this case the instantiation of the fundamental law would be:

Organisms whose seeds have traits that allow them to glide in the air spread their seeds better, thus improving their success in differential reproduction.

It is also interesting to point that in Darwin himself we can find adaptations presented at a wider level than that of the individual:

> "[...] it may be believed that under certain circumstances individual differences in the curvature or length of the proboscis, &c., too slight to be appreciated by us, might profit a bee or other insect, so that certain individuals would be able to obtain their food more quickly than others; and thus the communities to which they belonged would flourish and throw off many swarms inheriting the same peculiarities" (Darwin 1872, pp. 74-75).

The law-like statement presupposed in the explanation would be the following:

Bees whose tongues have the most effective curvedness or length to collect nectar from certain flowers will improve the performance of the community they belong to, improving the differential reproductive success of that community.

All these explanations share the same structure and there is an interesting sense in which it can be held that they are part of the same theory, even though Darwin himself uses the expression "natural selection" in a more restricted sense on certain occasions. Only if we consider this wide theory of natural selection can we understand why, according to Darwin, the strongest reason to accept the theory is its unifying force.

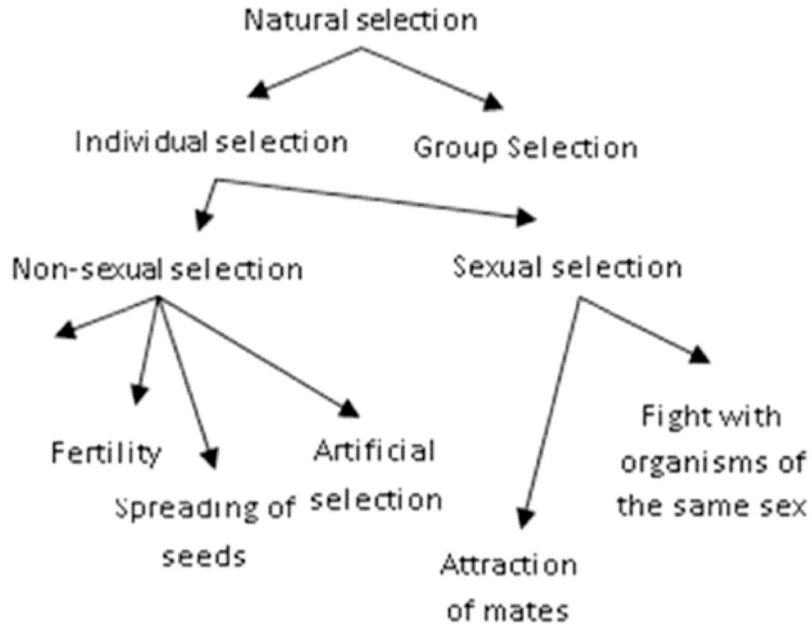

FIG. 1: THEORY-NET FOR THE THEORY OF NATURAL SELECTION

Structuralists use the expression "theory-net" (Balzer, Moulines & Sneed 1987) to refer to the web of special laws that arise by specialization from the fundamental law and that constitute the most usual sense that in practice is given to the term "theory". The theory-net based on the different specializations of *aptitude* would look like figure 1.

Nature of the variation

The fundamental law of the theory of natural selection as has been presented in the previous section presupposes the existence of the variation. In a given population it is possible to explain the differences in reproductive success only if there are differences regarding traits of the organisms from such population that imply differences in aptitude. But, as can be seen in the texts from Darwin reproduce above, he says nothing about the origin or the causes of the variation. That is because he just ignored them:

> "Our ignorance of the laws of variation is profound. Not in one case out of a hundred can we pretend to assign any reason why this or that part differs, more or less, from the same part in the parents. But whenever we have the means of instituting a comparison, the same laws appear to have acted in producing the lesser differences between varieties of the same species, and the greater differences between species of the same genus" (Darwin 1859: 167).

Darwin was aware that he did not know the causes of variation. He only knew some empirical generalizations regarding them. The theory of natural selection was born independently from any theory about the causes of variation and its behavior, building upon the empirical fact of phenotypical variation. Did Darwin consider that the variation over which natural selection operated ought to be "blind" in the sense explained above? In countless places in the Origin of Species Darwin subscribes to what Mayr calls "soft inheritance" (Mayr 1982: 687-698). In a characterization anachronical for its application to Darwin's ideas, Mayr describes the belief in soft inheritance as the belief that genetic material was flexible. Of course, Darwin did not speak of genetic material in any way, but there are two beliefs of his that, according to Mayr, would imply the belief in soft inheritance. The belief in the direct effects of the environment and in the effects of use and disuse as causes of evolution. Regarding the direct influence of the environment we can find many passages in which Darwin holds that the direct action of the environment could have had some influence in the production of the races. For instance:

> "Some little effect may, perhaps, be attributed to the direct action of the external conditions of life" (Darwin 1859: 29).

or,

> "How much direct effect difference of climate, food, &c., produces on any being is extremely doubtful. My impression is, that the effect is extremely small in the case of animals, but perhaps rather more in that of plants" (Darwin 1859: 132).

As for the effects of use and disuse,

> "I think there can be little doubt that use in our domestic animals strengthens and enlarges certain parts, and disuse diminishes them; and that such modifications are inherited" (Darwin 1859: 134).

These passages show that Darwin thought that the direct instruction of the environment and the two laws from Lamarck mentioned before work as mechanisms alternative to natural selection. This is possible only because Darwin, in effect, thought of inheritance as "soft". But these mechanisms also have an interesting role regarding our current issue. Darwin is explicit in holding that those mechanisms worked, in addition, as causes of the variation over which natural selection operated:

> "I may add, that when under nature the conditions of life do change, variations and reversions of character probably do occur; but natural selection, as will hereafter be explained, will determine how far the new characters thus arising shall be preserved" (Darwin 1859: 15).

Near the end of the chapter dedicated to the laws of variation, after revisiting all the possible causes of variations, among which we find use and disuse and the influence of the environment:

> "Whatever the cause may be of each slight difference in the offspring from their parents—and a cause for each must exist—it is the steady accumulation, through natural selection, of such differences, when beneficial to the individual, that gives rise to all the more important modifications of structure, by which the innumerable beings on the face of this earth are enabled to struggle with each other, and the best adapted to survive" (Darwin 1859: 170).

Darwin, therefore, believed that the variations over which the selective mechanisms ran could be caused by direct influence of the environment and as a result of use and disuse. That variations might be caused by the influence of environmental conditions goes against Campbell's requisite (2) mentioned before that variation should be blind. That they might be caused by use and disuse is incompatible with requisites (3) and (4). That is, the variation for Darwin did not need to be blind in Campbell's sense (Hodgson 2001: 103-105; Kronfeldner 2007: 499).

The additional explanatory power of the Darwinian theory of natural selection consisted precisely in that it could work on variations that did not depend on habit and were, therefore, more or less spontaneous. The force of the theory lay in that the variation might not arise as a response to the demands of the environment. But it did not have to be like this in all cases. The issue of it being impossible to distinguish between Lamarckian and Darwinian mechanisms if variation is blind can, I think, be eluded by paying attention to the complexity of the explanation. On the other hand, there is no risk either of the theory of natural selection becoming vacuous or irrefutable by making it more general than the version that works only on blind variations, since each of its parts is independently testable. If we claim, for instance, that fowl with tails with certain characteristics are more attractive to females and, therefore, get to mate more, improving their reproductive success, each of these claims can be tested by empirical means, with absolute independence of the nature of the variation. Finally, if the variation was generated by non-blind mechanisms, this does not make natural selection irrelevant, since the particular variation obtained non-blindly might still improve or worsen the reproductive success of the organism (Dawkins 1986: 300).

Darwinian cultural adaptions

As we saw, Darwin thought it possible for natural selection to work over directed variations. This makes it conceptually possible for the theory he devised to be applied to cultural evolution. But, in addition, he expressed, even if on few occasions, his openness to applying his theory without the need that variations be transmitted through hereditary material (Lewens 2008):

> "Now, if some one man in a tribe, more sagacious than the others, invented a new snare or weapon, or other means of attack or defence, the plainest self-interest, without the assistance of much reasoning power, would prompt the other members to imitate him; and all would thus profit. The habitual practice of each new art must likewise in some slight degree strengthen the intellect. If the new invention were an important one, the tribe would increase in number, spread, and supplant other tribes" (Darwin 1874: 129).

In this case the trait arises as a response to a need, that is, by a directed non-blind variation, and the inheritance mechanism is not genetics, but imitation. We can find, however, the same structure as in the cases considered. There is a trait, in this case a cultural one, the new technological artifact, and there is an improvement in the aptitude of the tribe that will lead to an improvement in its reproductive success. It is a case of selection at group level.

Another interesting case is the following:

> "We see variability in every tongue, and new words are continually cropping up; but as there is a limit to the powers of the memory, single words, like whole languages, gradually become extinct. As Max Müller has well remarked:—"A struggle for life is constantly going on amongst the words and grammatical forms in each language. The better, the shorter, the easier forms are constantly gaining the upper hand, and they owe their success to their own inherent virtue." To these more important causes of the survival of certain words, mere novelty and fashion may be added; for there is in the mind of man a strong love for slight changes in all things. The survival or preservation of certain favoured words in the struggle for existence is natural selection" (Darwin 1874: 90-91).

In this case, the theory of natural selection does not even apply to living organisms, but, in a way that seems to anticipate memetics, to forms of language. The Darwinian theory of natural selection, in addition to being consistent with it, was applied to cultural evolution by Darwin himself.

Conclusion

I have presented a sketch of the fundamental law of the theory of natural selection as it was conceived by Darwin and of the several ways in which it was applied. I have tried to show that if the claim that cultural evolution is Lamarckian only holds that cultural traits obtained as a response to demands of the environment are transmitted to offspring through imitation or more complex forms of learning, that does not imply that natural selection cannot play an explanatory role regarding the evolution of traits of this kind. This supposed inconsistency is based on the usually held idea that natural selection works over a blind variation and, as we saw, it is possible to hold versions of natural selection that only require variation. One example is the theory as it was conceived by Darwin. Of course, one can hold, as many epistemologists do (such as Popper or Campbell), that cultural evolution is not Lamarckian in this sense. But if it was, that does not imply that it cannot be studied in a Darwinian way. The use of the phrase "cultural evolution, in opposition to biological evolution, is Lamarckian" as it is usually used to attack Darwinian studies of cultural evolution, does not have any conceptual ground at all.

References

Álvarez, Juan Ramón 2009. De aquel Darwin tan singular al Darwinismo Universal: La problemática naturalización de las ciencias de la cultura. , *Ludus Vitalis* XVII (32):307-326

Balzer, Wolfgang, Carlos Ulises Moulines, and Joseph D. Sneed 1987. *An architectonic for science: the structuralist program.* Dordrecht, Lancaster, Reidel

Boyd, Robert, and Peter J. Richerson 1985. *Culture and the Evolutionary Process* Chicago, University of Chicago Press

Campbell, Donald T. 1974. Evolutionary Epistemology, in Paul Arthur Schilpp (ed.), *The Philosophy of Karl Popper*, La Salle, Illinois, Open Court, 413-463

Darwin, Charles 1844. Essay of 1844, in Francis Darwin (ed.), *The foundations of The origin of species. Two essays written in 1842 and 1844*, Cambridge, University Press,

Darwin, Charles 1859. *On the origin of species by means of natural selection.* London, John Murray

Darwin, Charles 1872. *The origin of species, 6th ed.* London, John Murray

Darwin, Charles 1874. *The Descent of Man and Selection in Relation to Sex.* 2th ed. London, John Murray

Dawkins, Richard 1986. *The blind watchmaker.* Harlow, Longman Scientific & Technical

Fog, Agner 1999. *Cultural Selection.* Dorbrecht, Kluwer

Fracchia, Joseph, and Richard C. Lewontin 2005. The Price of Metaphor, *History and Theory* 44:14-29

Gomila, Antoni 2009. El peligroso Darwin y las Ciencias Sociales, *Ludus Vitalis* XVII (32):339-352

Gould, Stephen Jay 1980. *The Panda's Thumb.* New York, Norton

Gould, Stephen Jay 1993. *Eight little piggies. Reflections in natural history*, W.W. Norton

Gould, Stephen Jay 1996. *Full House. The Spread of Excellence from Plato to Darwin*. New York, Harmony Books

Hahn, Hans, Otto Neurath, and Rudolf Carnap 1929. *Wissenschaftliche Weltauffassung: Der Wiener Kreis*. Wien, Artur Wolf Verlag (Versión castellana de Pablo Lorenzano, "La concepción científica del mundo: el Círculo de Viena", en *Redes*, Vol. 9, N° 18, Junio de 2002).

Hodgson, Geoffrey M. 2001. Is Social Evolution Lamarckian or Darwinian?, in John Laurent and John Nightingale (eds.), *Darwinism and Evolutionary Economics*, Cheltenham, Northampton, MA, Edward Elgar, 87-120

Kronfeldner, Maria E. 2007. Is cultural evolution Lamarckian, *Biology and Philosophy* 22 (4):493-512

Lamarck 1809. *Philosophie zoologique*. Bruxelles, Culture et Civilisation

Lewens, Tim 2008. Cultural Evolution, in Edward N. Zalta (ed.), *The Stanford Encyclopedia of Philosophy*.

Mayr, E. 1982. *The Growth of Biological Thought*. Cambridge, MA, Harvard University Press

Medawar, Peter 1960. *The future of man: the BBC lectures 1959*. London, Methuen

Popper, Karl 1974. *Unended Quest: An Intellectual Autobiography*. La Salle, Open Court (versión castellana de Carmen García Trevijano, *Búsqueda sin término: una autobiografía intelectual*, Madrid: Tecnos, 1977).

Richards, Robert J. 1987. *Darwin and the emergence of evolutionary theories of mind and behavior*. Chicago, University of Chicago Press

Toulmin, Stephen, E. 1972. *Human Understanding: The Collective Use and Evolution of Concepts*. Vol. I. Princeton, Princeton University Press

THEORY OF CLASSIFICATION AND TAXONOMICAL SCHOOLS: A SYNTHESIS FOR ARCHAEOLOGY

Daniel García Rivero

Departamento de Prehistoria y Arqueología, Universidad de Sevilla, Facultad de Geografía e Historia, C/ María de Padilla s/n, 41004, Sevilla, España,
e-mail: garciarivero@us.es Teléfonos: (0034) 954556906 / 679879269 Fax: (0034) 954559920

Abstract

Taxonomy works forming groups of specimens or of any kind of object on the basis of a particular series of criteria. Its use can be extended to all areas of knowledge of historical nature in which variation and diversity are constantly produced through time and space. Although taxonomy has been applied to anthropology and archaeology for a long time, most of the methodological achievements have taken place in evolutionary biology.

The main aim of this paper is to introduce researchers in social sciences, especially archaeologists, to the fundamental theory underlying all of the taxonomical reconstructions and classifications, as well as to present the historiography and methodological development of this research area.

However, the main body of the paper is concerned with the presentation of the different philosophical schools of taxonomy: phenetics, cladistics and evolutionary taxonomy. Due to the important implications that this has for our approaches it is essential to define each one clearly and to undertake an explicit comparative analysis between them within an archaeological framework.

KEYWORDS: ARCHAEOLOGICAL THEORY, TAXONOMY, CLASSIFICATION, PHENETICS, CLADISTICS, EVOLUTIONARY TAXONOMY.

Resumen

La Taxonomía trabaja formando grupos de especimenes u objetos según una serie de criterios particulares. Su aplicación puede extenderse a todas aquellas áreas de conocimiento de naturalezas históricas donde se produce constantemente variación y diversidad a lo largo del tiempo y del espacio. Aunque la Taxonomía se ha aplicado a la Antropología y a la Arqueología desde hace mucho tiempo, la mayoría de logros metodológicos ha tenido lugar en la Biología evolutiva.

El principal objetivo de este artículo es introducir a investigadores de ciencias sociales, especialmente arqueólogos, a los fundamentos teóricos subyacentes a toda reconstrucción taxonómica y tarea de clasificación, así como exponer la historiografía y el desarrollo metodológico de esta área de investigación.

No obstante, la parte principal del trabajo concierne a la presentación de las diferentes escuelas de la Taxonomía: Fenética, Cladística y Taxonomía evolutiva. Debido a las importantes implicaciones que esto tiene en nuestras aproximaciones, es esencial definir claramente cada una de ellas, así como establecer un análisis comparativo explícito entre las mismas dentro de una discusión de naturaleza arqueológica.

PALABRAS CLAVE: TEORÍA ARQUEOLÓGICA, TAXONOMÍA, CLASIFICACIÓN, FENÉTICA, CLADÍSTICA, TAXONOMÍA EVOLUTIVA.

> ..The most impressive aspect of the world of life is its diversity and the uniqueness of its components. … Wherever we look in nature, we find uniqueness, and uniqueness means diversity. To sort all this and determine its nature is the task of taxonomy. (Mayr and Ashlock 1991: 1).

A brief historiographical introduction to taxonomy

The ordering and grouping of all kinds of beings and objects has been one of the main tasks since the dawn of scientific knowledge, and even long before. These tasks were not only taken up in important works and efforts among the classical thinkers, the scholars of the Renaissance and the Enlightenment, and the whole of the contemporary scientific community. Indeed, since Prehistoric times, the human perception has classified numerous animal and vegetal species, as well as a wide spectrum of stones, minerals and other objects.

The response to this incipient tendency to group and order individuals and groups may have been more complex than we often imagine, but was undoubtedly linked, on one hand, to the interest and the knowledge of the world that surrounds (in this case) the human species and, on the other hand, to an attempt to synthesise and understand the variability present within nature. Thus perhaps the main reason for classification is the creation of groups within large sets of specimens or objects that enable us to synthesise information and therefore to better understand the world that surrounds us. On the scientific level, classification makes the results of a study more intelligible and is therefore an efficient factor in the transmission of information and knowledge.

Understood as a tool of scientific research, taxonomy emerged in the 17th century and became established as a discipline of knowledge in the 18th century with the works of C. Linnaeus. His Sistema Naturae included a hierarchical classification with a number of degrees of inclusion, the names of which remain today (kingdom, class, order, family, genus and species). Since Linnaeus himself conceived Nature as a static and immutable divine design, the main criterion in his approach to the grouping of individuals was similarity. Thus the specimens or groups of specimens that share most similarities belong closer to each other in the classification. This methodology, still valid today, is known by the term phenetics (although further detail will be given below, for now it is useful to note that phenetics uses exclusively the criteria of similarity between individuals to establish taxonomic groups, without considering the implications of evolution and the laws of inheritance, and therefore without distinguishing between the different types or causes of similarity).

However, the main work of Charles R. Darwin (1859) in the following century would imply the consideration of a new perspective in the processes of elaboration of classifications. Indeed the idea of the transformism of species, as elements that change over time, and the idea of inheritance already existed in the work of J.B. Lamarck (1809). Darwin based his theory of descent with modification on the Linnaean taxonomic model but incorporated natural selection as the mechanism behind change. A quote taken from a recent study describes clearly though succinctly the idea that Darwin himself held on how individuals were to be grouped in the creation of taxonomic groups: «I believe that the arrangement of the groups within each class, in due subordination and relation to the other groups, must be strictly genealogical in order to be natural» (Darwin 1859: 420, quoted in O'Brien and Lyman 2005: 94). And this was well illustrated by the single figure included in his famous book (Darwin 1859, Diagram pp. 116-117).

This idea and its systematic application became consolidated during the 20th century. The neodarwinian synthesis –what Huxley (1942) referred to as the Modern Synthesis- enabled the epistemological and ontological realignment of the Linnaean taxonomic model with Darwinian evolutionary theory (O'Brien and Lyman 2005: 95).

The works of W. Hennig in the 1960s (e.g. 1965; 1966), though published in his native German language in the previous decade (1950), meant the consolidation of the phylogenetic systematic.

The approach known finally as cladistics considers the evolutionary laws and the characters linked exclusively to the relationships of descent in the creation of taxonomic groups.

However, the cladistic model was too strict in several ways for many taxonomists, as we shall see below. A further taxonomic approach thus was developed in the following decade, in good part due to E. Mayr (Mayr 1969; Mayr and Aschlock 1991), which would become known as evolutionary taxonomy. This approach displays various objections and controversies with respect to cladism, but effectively assumes the Darwinian phylogenetic ideas. For now, this third school may be defined as a fusion and synthesis of the phenetic approach based on distances and the exclusively genealogical cladistic approach.

These three schools of taxonomic thought –phenetics, cladistics and evolutionary taxonomy- remain valid today, although the latter two are by far the most common in the study of biology. They are not however as common in other fields of knowledge in which it is of key importance to establish groups within large sets of individuals, as is the case of those disciplines concerned with the study of culture and human behaviour, as are anthropology and archaeology. Most archaeologists continue to work exclusively with the criteria of how similar one find is to another. And little interest has been placed on the problem of the consideration of cultures and archaeological sets as independent entities. This was noted in relation to anthropology by F. Galton, Darwin's cousin, who remarked that the comparative studies of human adaptation are irrelevant if the possibility of different cultures sharing a common origin for the adaptative trait under study cannot be discarded (O'Brien et al. 2008: 39). In summary, it is of key importance to delve into greater detail and determine whether the traits shared by different taxa (cultures, populations, etc.), and that we use to group them together in our classifications, are shared because they stem from a common ancestor or for other reasons (adaptation, cultural borrowing, evolutionary convergence, etc.). As we shall see further below, phylogenetics is confronted with this issue. Fortunately, the last decade has witnessed the proliferation of anthropological and archaeological approaches using phylogenetic methods, leading to the publication of several monographic works on the subject (e.g. O'Brien and Lyman 2003; Mace, Holden and Shennan 2005; Lipo et al. 2006).

On the concept of species

The species is generally considered as the most basic unit or taxon of the hierarchical scale of biological classification. Therefore, the idea or unit of species is closely linked to taxonomy. Here we do not aspire at all to enter into the detail of the concept of species nor into the debate regarding this issue in the field of biology. However we believe it important to mention, though briefly (due to the space available) some aspects of high relevance. And by this we intend to promote a reflection on the issue, in the mind of the archaeologist, that is to centre the question on the units or types that we normally use in archaeology.

Not all taxonomists understand the same by species. Indeed, the concept of species is so plastic for biologists that only with difficulty, were it to be possible, could one find a definition to the liking of everyone. Currently there are a great number of different understandings of the term, from the twenty-two definitions recovered from the recent bibliography used by R.L. Mayden (1997) to the limited half a dozen acknowledged by most researchers. But since when and most of all why does this debate exist?

The debate has existed since the emergence of scientific taxonomy. C. Linnaeus understood species under the vision of a primeval stability that led him to affirm that: «There are as many species as the Infinite Being produced diverse forms in the beginning» (LINNAEUS 1735, quoted in Zimmer 2008: 74). J.B. Lamarck, in contrast, perceived the idea of their transformism:

> "...Cette définition est exacte; car tout individu jouissant de la vie, ressemble toujours, à très-peu près, à celui ou à ceux dont il provient. Mais on ajoute à cette définition, la supposition

que les individus qui composent une espèce ne varient jamais dans leur caractère spécifique, et que conséquemment l'espèce a une constance absolue dans la nature.C'est uniquement cette supposition que je me propose de combattre, parce que des preuves évidentes obtenues par l'observation, constatent qu'elle n'est pas fondée…"(Lamarck 1809: 54).

A brief quote of Darwin directs us towards the current debate on the concept of species: «I look at the term 'species' as one arbitrarily given, for the sake of convenience, to a set of individuals closely resembling each other» (Darwin 1859, quoted in Zimmer 2008: 74). The discussion on the concept of species intensified considerably throughout the 20th century, with the emergence of the Synthesis in the 1940s and with the development of the systematic phylogenetic methods in the 1950s and mainly 1960s.

With regard to the second part of the question formulated above, asking why this discussion exists, there are mainly three factors. First, the immense plasticity and diversity displayed by the living and fossil record generally makes difficult the definition and demarcation between species. Second, the ontological nature of the term itself, or, in other words, what the theoretical premise of what a species should be, causes variable responses. Is a species a discrete unit, a product of evolution, or on the contrary is it an analytical definition used to describe the variation and diversity of the organic record? The first perspective is known as a realist species and the latter as a nominalist species (Ridley 1996: 420). Third, the different approaches in different fields of research are also an influence. Indeed, the definition of species varies between taxonomists concerned with the study of mammals, birds, plants, or bacteria due to the differences in the intensity and in the processes and mechanisms of variation themselves, as well as to the different scales of analysis. The different characteristics or patterns of variation and speciation of the record are particularly important and have relevant implications such as the evolutionary rhythm of one or another species. Depending even on how one considers speciation (the process by which new types of organisms emerge from ancestral forms), there will be different approaches to what a species should be and, moreover, to how these species should be sought and established by means of the taxonomy of one or another field of knowledge.

In the following section, we will focus only on four definitions of species that are accepted by most taxonomists: typological-phenetic, biological-reproductive, phylogenetic and evolutionary. Other definitions of species commonly accepted are those based on concepts of cohesion, recognition or ecology. For further detail on the debate on the definition of species see (Hennig 1966; Mayr and Ashlock 1991; Ridley 1996; Mayden 1997; Wheeler and Meier 2000; and also, Coyne and Allen 2004, cited in Zimmer 2008).

First of all, the typological concept of species is based on the global similarity between specimens, which are normally grouped together following a previous taxon as an ideal type. It focuses essentially on morphological characteristics and does not take into account the evolutionary principles. This definition of species characterises the taxonomical school known as Phenetics; both (the definition and the school) still dominate in archaeological classifications.

The biological-reproductive definition of species is the most followed definition in Biology. Reproductive crossing with fertile offspring is the criterion that enables two individuals to be attributed to the same species (Dobzhansky 1937: 228-258; Mayr 1963: 89-109). It is therefore an atemporal –and non-dimensional- concept of species that is not applicable when time is to be considered. The biological definition is also problematic since is cannot be applied to asexual organisms, thus supposedly making it inapplicable to archaeological material culture.

In third place, the phylogenetic concept of species (sometimes referred to as hennigian) possesses numerous definitions, all of which are partly different (cf. Hennig 1965; 1966; Cracraft 1983; 1987; Meier and Willman 2000; Mishler and Theriot 2000; Wheeler and Platnick 2000). Despite their differences, as we shall see below, they possess a common taxonomical definition based on the presence of derived traits that are shared by two or more taxa in common with their immediate ancestor. It is a vertical and genealogical conception, with a temporal dimension. Indeed, these definitions of species dominate in the remaining taxonomical schools known as Cladistics and Evolutionary taxonomy. What is now known as Cladistics was previously named as Systematic phylogenetics, after the works of W. Hennig (1965; 1966). Currently, Cladistics and Evolutionary taxonomy are generally known as Phylogenetics.

Finally, the evolutionary conception of species can be traced at least to the works of G.G.Simpson (1951; 1961). The height of the discussion of the concept of species in relation to systematic phylogenetics, from the 1980's onwards, led many authors to use the evolutionary definition in their approaches (Wiley and Mayden 2000: 71-72). This definition minimises the importance of the characterisation of groups based on shared derived characters and considers other types of evidence such as the mode of reproduction, the genetic flow, and the geographical information. It requires the union of the phylogenetic methodological procedure with information on processes of coevolution, historical ecology, biogeography, speciation and palaeontology. This definition of species is found in the works of Evolutionary taxonomy.

On characters and units in taxonomy

One of the basic issues in taxonomy concerns the elements used to order individuals and groups. These sections of information that we extract from the objects under study are of great importance. Thus when we are undertaking the study of an assemblage with the aim to classify its components, it is always important to reflect on how and why to extract one or another section of the total information available. A thorough reflection, taking into account the questions that are to be approached (what is it we want to answer?) and the scales of analysis, will be productive.

These pieces of information are what the taxonomist generally calls characters. They may be binary, that is with two possible measures or answers; for instance, the states of presence or absence of handles on pottery. But they may also be multistate, for example the colour of the surface of pottery. This duo of terms (character and state), so useful in coding and analysing information, can also be found in the archaeological literature for example as variable and attribute (Adams 1988: 47-48). However, a widely used concept that covers the etymological needs of different fields of knowledge is that of trait.

There are different types of characters or traits. The most commonly used, in early studies and in recent work (with the exception of specific fields such as Molecular Biology), are the traits corresponding to morphological aspects. This information refers to shape, size, number or quantity, etc., in short, to all of the details that may be observed through the visual analysis of an object as a tangible physical entity.

Another type, often used since the emergence of Genetics, is the molecular trait or DNA. The molecular information held by the genomes has recently received much interest in taxonomical studies, mainly phylogenetic (e.g. Yang and Goldman 2008). An important part of this information is transmitted from one gene to another during the chromosomic replicative processes. That is, this information enables us to use the chains of genomes to reconstruct their evolution and to group the specimens over time. Recently, some information of this kind has emerged in paleonthology, although its use in archaeology is still clearly minority.

Lastly, a final type is given by the behavioural traits that are fundamental in studies of ethological nature. Naturally, these attributes can be easily observed in living populations but are difficult to recover for extinct groups. However, these behavioural traits may ultimately be inferred indirectly from the material culture, through the intermediary of morphological or molecular traits.

A detailed reflection on the characters and traits themselves, as measurable units that enable the scientific process, has important ontological implications. Archaeological studies that have gone into the detail of this question (e.g. Dunnell 1971; O'Brien and Lyman 2002) underline the versatility of the units in taxonomy, depending whether or not they are given a theoretical context.

In this sense, the units of measure may be divided into two groups: descriptive units and analytical units, also named theoretical units by some authors (O'Brien and Lyman 2002: 41), which are created for a specific purpose. On one hand, the units may be used descriptively. Many empirical studies of culture history, among other tendencies, describe archaeological objects as an end in itself. However, these descriptions serve little to answer specific questions (of the type why and how) regarding the processes or other interpretative aspects involved. Such descriptions do, however, constitute excellent bodies of data available for later studies. On the other hand, the units may be analytical in nature. These units correspond to carefully selected characters which, by means of any given methodology, enable us to answer a series of specific questions, formulated beforehand (even prior to the collection of data). Of course, the states or attributes of each one of the characters must be designed in function of the questions to be asked. Once the corresponding experimentations have been carried out, by means of any given method, the analytical traits will enable us (at least) to reject or maintain the working hypothesis, thus answering our question. The creation of a set of appropriate units for a specific purpose, process that we could call systematic, provides the necessary relationship between the theory and the empirical world (O'Brien and Lyman 2003b: 15).

The units may further be derived extensionally or intentionally. In the first case, the units are defined on the basis of the observable attributes that are shared by the individuals. And they do not require an explicit theoretical basis. The criteria that constitute the units are thus based on the observable traits of the specimens currently classified within a group. This type of unit is thus influenced by the provisional nature of the sample at each point in time. This may cause that, faced with the discovery of a new specimen, the classification must be reviewed and reconstructed. The second case, in contrast, requires the theoretical specification of the necessary and sufficient conditions to be met for a specimen to be classified within a particular unit (O'Brien and Lyman 2002: 42).

On the types of classification

Frequently there exists some confusion over some aspects of taxonomy, not only in archaeology but even among experienced taxonomists, especially regarding the relationships between units or the forms of classification (Dunnell 1971: 68-69).

Classifications consist of sets of related characters, and thus differ according to the relationships between the distinctive constituent traits, since some may be of greater importance than others, i.e., of greater weight. The distinctive traits necessary for a specimen to be considered of a particular class are named differentia (Stearn 1959; cited in O'Brien and Lyman 2003a: 25) or significata. The specimen itself (individuals that possess significata of a particular class) is the denotata (Dunnell 1971: 45 and 70).

But as we have seen above, the way in which the traits themselves are considered is important. These may be dealt with in two main ways with regards to the relationship between them. On one hand, they may be considered as equivalent, with the same weight, and associated to analogous attributes in other objects (intersection). On the other hand, they may be structured hierarchically in such a way that they each possess a different weight and are thus associated by inference (inclusion). This results in two types of completely different classification: paradigmatic and taxonomic, respectively (Dunnell 1971: 70).

The paradigmatic classification (Figure 1a) creates groups using dimensions of traits. A dimension is a set of mutually exclusive alternative traits, and they differ according to the final product obtained from the possible combinations of dimensions (Dunnell 1971: 71). An essential characteristic of the paradigmatic classification is that it presents no ambiguities, in terms of its internal structure and in the creation of groups of phenomena. It is objective since the classes produced are established with the same set and weighting of traits. Any paradigmatic classification may be expanded infinitely.

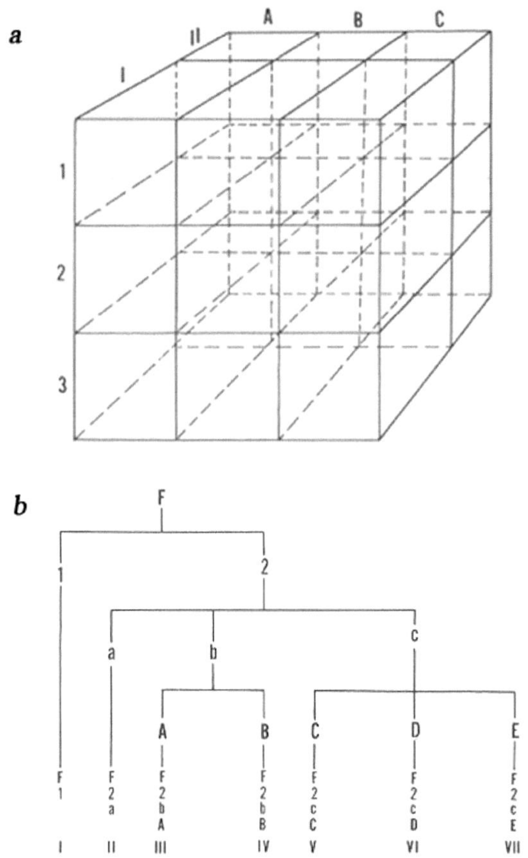

FIGURE 1: GRAPHICAL REPRESENTATIONS OF THE PARADIGMATIC (A) AND TAXONOMIC CLASSIFICATIONS (B). FIGURE ELABORATED AFTER (DUNNELL 1971, FIGURES 4 AND 6, RESPECTIVELY).

Taxonomic classification (Figure 1b), in contrast, assumes from the outset a hierarchical order of the traits, based on the premise of the Linnaean taxonomic model of Nature. The family-based hierarchical structure of the taxonomy is based on distinctive non-dimensional forms. The traits are mutually exclusive and are not permutable, since the position of each one of the traits is not arbitrary.

If we compare both types of classification, the differences mentioned above are evident. The choice of one or another type will depend on diverse factors. But it is noteworthy that if the initial assumptions (due to the intuition of the researcher and the use of the hierarchical model) can be tested through experimentation, and suggest a high degree of probability, then the taxonomic classification is ideal. However, if the classes must be known prior to the taxonomic construction, as is frequent in studies of phenomena that remain little known (as may be the case in many archaeological issues), and the assumptions must remain static (without taking a risk that implies subjectivity and ambiguity), then paradigmatic classification is preferable (Dunnell 1971: 82-84).

It is likely that the exclusion of such important aspects as those outlined above has caused confusion among classifications and researchers alike. In the North American archaeological context, the discussion over the creation of archaeological types goes back to the first half of the past century.

On the philosophical schools of classification

The work on classification carried out up until the present moment can be grouped basically into three philosophical schools. The first distinction to consider is whether the classification takes into account the principles of evolution by descent with modification. If the answer is no, that is if the classification does not assume these principles, then the approaches to classification may all be restricted to a school named phenetics. If, on the contrary, the evolutionary principles are accepted, then the classifications may belong to two theoretical schools: cladistics and evolutionary taxonomy. In the following sections we shall characterise each one of these schools. Each presentation includes the main principles and objections and some notes on their application. This theoretical section, inevitably, belongs to a general scientific field in which we find definitions and examples drawn from different areas of knowledge, mainly biology, although in so far as possible we shall attempt to illustrate some points with archaeological examples.

Phenetics

The term phenetics derives from phene, a concept that refers to any trait or characteristic of an organism. The phene alludes to the units of the phenotype in the same way as the gene refers to the units of the genome.

Phenetics is based fundamentally on the classification according to the general similarity between the specimens. On the basis of particular traits, some specimens are more similar to each other than they

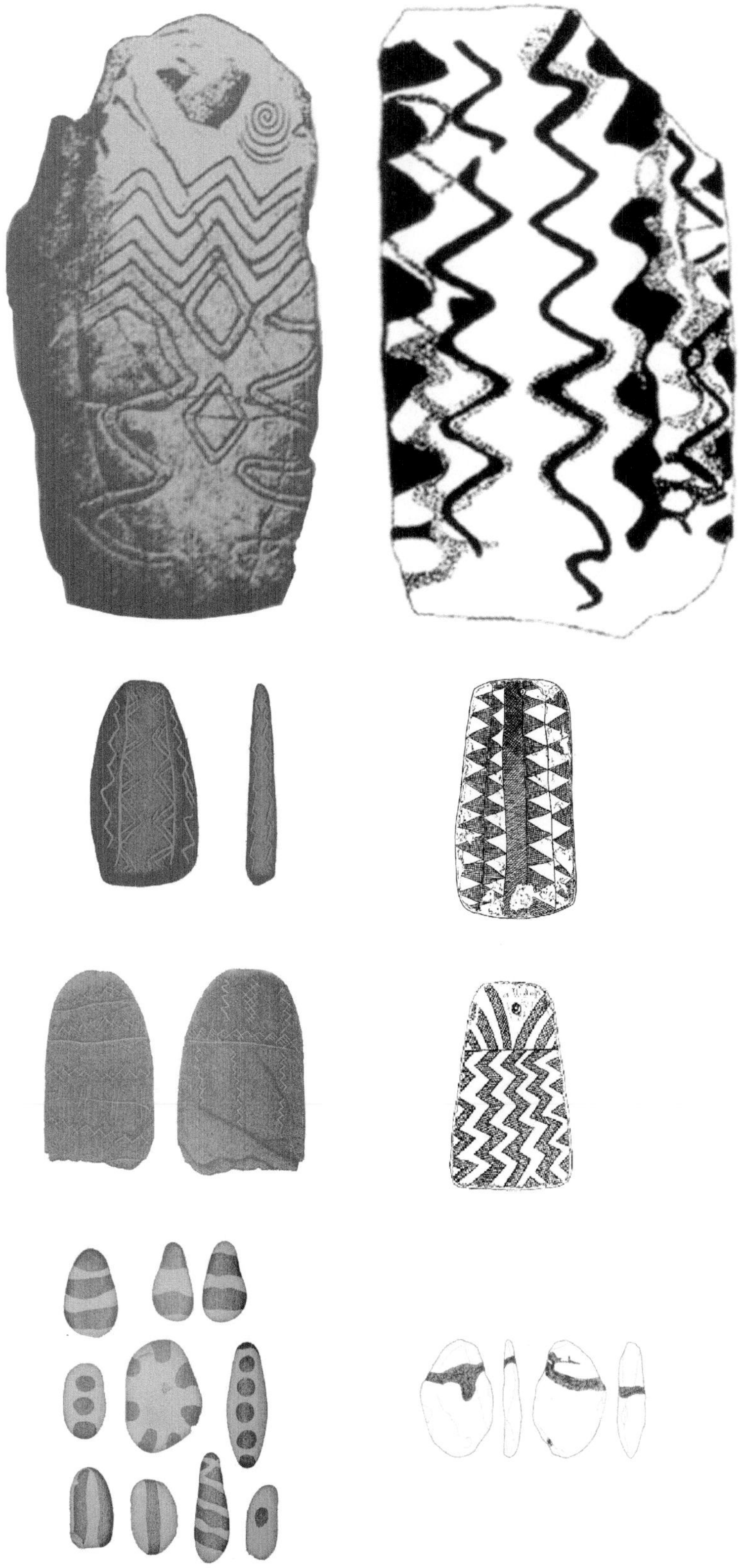

FIGURE 2: THE DECORATION OF MEGALITHIC UPRIGHTS AND OF THE IBERIAN 'PLAQUE IDOLS' ARE VERY SIMILAR – ALMOST IDENTICAL- TO THE MOTIFS FIGURING ON SPATIAL AND TERRITORIAL MARKERS AND ON OTHER PLAQUE-SHAPED ARTEFACTS IN NORTHAMERICAN INDIGENOUS SOCIETIES. THE SAME SIMILARITIES ARE FOUND WHEN COMPARING THE AMERICAN PAINTED RIVER PEBBLES WITH THOSE OF THE CANTABRIAN AZILIAN PERIOD. THE LEFT HAND COLUMN CORRESPONDS TO MATERIALS FROM INDIGENOUS TRIBES AND THE RIGHT HAND COLUMN IS FORMED BY PREHISTORIC MATERIALS FROM THE IBERIAN PENINSULA. (THE ELEMENTS ARE NOT TO SCALE). FIGURE ELABORATED AFTER (CARPENTER AND SCHUSTER 1986; BREUIL 1933-1935; LILLIOS 2004; BUENO ET AL. 2008).

are to other specimens. If we calculate the shared characters (similarities) and the variable characters (differences) between the individuals under study, it is possible to establish a concrete measurement that will inform us about the dual relationship between two specimens. And in a set we will obtain concrete measurements that inform us about the relationships between all of the specimens. These measurements effectively tell us the distance between each pair of specimens and the distances between all of the specimens. A previous step to these calculations is the construction of a matrix of distances that reflects all of the vectors that relate all of the cases.

Let us suppose that we must classify an assemblage of prehistoric pottery vessels. A first look at the assemblage will enable us to establish different groups based on the global similarities. Thus we would probably group the vessels mainly according to their form. However, we would probably suggest a number of different groups, or at least we would redefine the initial groups, if as well as the form of the vessels we considered their sizes, surface colours, techniques and decorative motives, or any other characters.

Thus, it is argued by phenetics that the higher the number of traits considered, the more precise and correct the out coming classification. The apogee of computer-based resources over the past decades has enabled greater possibilities of calculation. Indeed, the studies began to include such a high number of characters that the approaches relied on an ever greater use of statistical techniques and methods. This approach thus became known as numerical taxonomy (Sokal and Sneath 1963). Once faced with such a great quantity of data (characters, states and specimens) various multivariate methods of analysis were introduced, with the aim to synthesise the information and determine the contribution of diverse factors to the global set. Some of these procedures are factorial analysis, conglomerate analysis, principal component analysis, principal coordinate analysis, correspondence analysis, etc. The graphical representations obtained may generally be named phenograms, although there are a number of specific versions such as clusters, conglomerates and dendrograms.

According to some specialists, phenetics is efficient when applied to large groups with relatively similar specimens (Sokal and Sneath 1963; Jones and Gray 1994: 12). However, there are numerous objections. Some are so serious and relevant that the phenetic analysis was abandoned by most taxonomists once alternative approaches were developed. Thus, from the 1960s and 70s onwards, practically all taxonomist biologists left phenetics for other schools that took into account the evolutionary principles. This did not occur, and is still barely occurring, in the social sciences concerned with the study of humans.

The first objection is effectively that phenetics does not take into account the evolutionary history of the organisms or objects under study, and ignores the relationships of descent. And this implies serious theoretical and methodological flaws. Two processes or phenomena illustrate this explicitly: parallelism and evolutionary convergence.

Firstly, a parallelism is a trait that emerges by chance in two different places and times, lacking any phylogenetic connections, or what is the same, lacking a common ancestor. For example, Figure 2 compares some elements of the archaeological record of indigenous North American tribes with others from the Late Prehistory of the Iberian Peninsula. However, hardly anyone would defend that both material manifestations are similar, or almost identical, because they share a close common ancestor or because both populations and ethnic groups exchanged religions or symbolic universes. Rather it is an example of a random phenomenon that has led to a parallelism between two completely isolated populations and cultures, both geographically (on either side of the Atlantic Ocean) and in time. Now, in this hypothetical example we know the spatial and temporal discretion of the two societies taken as examples, but what happens when we are dealing with archaeological materials from a same continent and period? Distinguishing whether the traits are similar due to parallelisms or to links of common descent is practically impossible, or at best ambiguous. Thus the criterion of resemblance or global similarity of the phenetists has been strongly criticized.

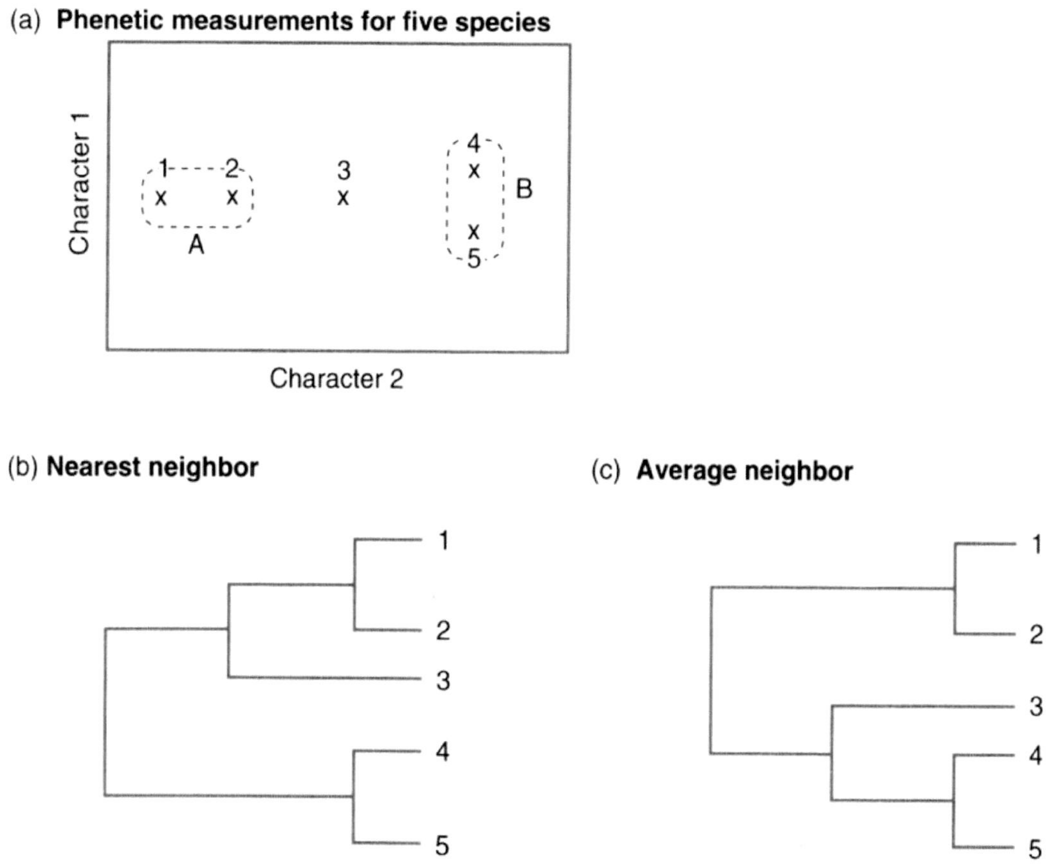

FIGURE 3: EXAMPLES OF TECHNICAL AND METHODOLOGICAL AMBIGUITIES IN PHENETICS. THIS ILLUSTRATES THE DILEMMA CAUSED BY THE CLASSIFICATION OF THE SPECIES NUMBER 3, SINCE IT FINDS ITSELF RIGHT IN THE CENTRE OF THE DISTANCE THAT SEPARATES THE SETS A AND B. DEPENDING ON THE METHOD OF GROUPING THAT IS USED, IT WILL BE CLASSIFIED SUBJECTIVELY IN ONE OF THE OTHER SET. FIGURE ELABORATED AFTER (RIDLEY 1996, FIGURE 14.4).

Secondly, an evolutionary convergence refers to a trait that is found in two different places and times, lacking any phylogenetic connection, but whose emergence is due to similar processes of adaptation in both cases. Effectively it is a phenomenon similar to the parallelism described above, with the difference that in the case of an evolutionary convergence the main reason for the development of a particular trait on several occasions responds to similar selective pressures in different environments. Moreover, it occurs (apparently) more often than parallelism. An illustrative example is the trait fur or feather colour in different animals. The snow partridge and the snowy owl, just like the polar bear, possess white feathers and fur, as the result of the selective pressures of a snowy environment. The external colour of the organisms of these species has been modelled by the natural selection of populations that arrived and reproduced, over generations, in Arctic regions. This example demonstrates the emergence or modification of traits towards identical states in different animal species that share no recent phylogenetic history.

With respect to some archaeological assemblages, often part of poorly known phenomena, this warns us of the seriousness of this type of issue and should lead us to question the classificatory approach that currently dominates in archaeology.

Furthermore, the objections to phenetics are not only philosophical, as those outlined above. Phenetics also suffers purely technical and methodological problems. For example, some specimens may often belong right in the middle of the distance between two groups (Figure 3). The placing of these intermediate specimens in one or the other group is based on ambiguous and subjective decisions, particularly given that we know that different and contradictory results will be reached depending on the measures of distance that we use.

Cladistics

The term cladistics derives from the Greek concept of klados, meaning branch. The coinage of this term for this method refers to the process of speciation by means of the model of divergence that it assumes, and that is metaphorically illustrated by the trunk and branches of a tree.

The phylogenetic systematic, currently known more specifically as cladistics, emerged as an alternative taxonomical philosophical school to Phenetics, with the intention to eradicate the objections outlined above. In the first half of the 20th century some attempts were made to innovate the taxonomical approach, for instance by the work of W. Zimmerman (1931, cf. Donoghue and Kadereit 1992). However, it was the work of W. Hennig (1950; 1965; 1966) that put forward a coherent structuration in terms of the systematization and objectivity of the concepts and criteria of analysis involved. The criterion of resemblance and similarity was relegated and the principles of inheritance became the unique criteria upon which to group assemblages of specimens.

Effectively, Hennig formalised and consolidated the methodological process of systematic phylogenetics, but the idea that the genealogical principles must define the reconstruction and classification of organisms and other types of entities (to obtain correct classifications) had been in the minds of previous researchers. Even before the naturalists, some linguists sensed that the historical languages must have come from ancestral languages. For example, the Dutchman M. Z. van Boxhorn, who recognised in 1643 the Indo-European linguistic family, conceived languages as organic systems of grammatical regularities and suggested that the similarity between the Indo-European languages may be due to the existence of a primitive common language (van Driem 2001: 159). W. Jones (1786) continued with this idea and suggested that the similarity between Sanskrit and the European languages, such as Greek and Latin, was a result of their emergence from a fairly recent single ancestral language, now known as Proto-Indo-European. Barely five years after the publication of the first edition of the Origin of Species, a genealogical study of the Indo-European languages was published by A. Schleicher (1863). This work offers a classification of the evolution of some Indo-European languages, in the purest style of the current phylogenetic graphical representations (Schleicher 1863, cited in Forster and Renfrew 2006, Figure I.2.). This evolutionary conception in the linguistic field was not limited to the study of the evolution of languages. The idea that the systematic study of the variants (changing elements) in the traditions of manuscripts is genealogical in nature was claimed in the first half of the 19th century by students and analysts of manuscripts (stemmatic analysis) such as K. Lachmann and C. Gottlieb Zumpt.

But let us focus here on biology because it was in this field that the phylogenetic methodological systematization took place, and because we would otherwise stray from the main theme of this paper. In the first section of this study, we referred to a short quote by Darwin regarding the basic fundamentals of classification. It is also worth recalling a known quote by Simpson, which exemplifies perfectly the difference between the phenetic and the phylogenetic philosophical perceptions:

Here it is necessary again to emphasize the distinction between definition and the evidence that the definition is met. We propose to define taxonomic categories in evolutionary and to the largest extent phylogenetic terms, but to use evidence that is almost entirely nonphylogenetic when taken as individual observation .... The well-known example of monozygotic ("identical") twins .... The individuals in question are not twins because they are similar but, quite the contrary, are similar because they are twins. Precisely so, individuals do not belong in the same taxon because they are similar, but they are similar because they belong in the same taxon. (Simpson 1961: 69).

Cladistic classification reflects the evolutionary history of the organisms or objects under study. It assumes a divergent model of evolution, through which two new species generally descend from a common ancestor (cladogenesis). The principle of divergence explains why phylogenies are hierarchical, in the same way as the Linnaean model. The traits that enable the reconstruction of the phylogenies of species and more inclusive groups must be reduced to those that display specific evidence of ancestor-descendent relationships. It is thus argued that not all traits may be necessary in the construction of classifications.

For this reason, an initial distinction is made between analogous and homologous characters (Figure 4). The first are characters that display the same state in different species when there is no evidence of genealogy between them. For instance, the cases that respond to parallelisms and evolutionary convergences. Homologous characters, in contrast, are those shared by two phylogenetically related species (ancestor-descendent relationships). In this case, the two species share the homologous character because they descend from a same common ancestor that already possessed that character. It is, however, possible that many other species exist between the common ancestor and the two species under study. Were this to be the case, the homologous trait may possibly be of use at an even higher inclusive level (family, genus) but may not be at all useful for distinctions on a smaller scale, since it would presumably be present in many different species. For this purpose, homologous characters can further be defined as ancestral and derived. The former correspond to the technical term plesiomorphy and the latter to the term apomorphy. That an attribute is considered as plesiomorphic or apomorphic depends on the level of inclusion (in terms of taxa) of the analysis itself. An ancestral homologous character, that illustrates didactically the issue, may be the presence of a vertebral column in the grouping between vertebrate animals or between mammals. Given that all species of vertebrates and mammals display a vertebral column, this trait would be of no use (a non-informative character). A derived homologous character, in contrast, may be for example the presence of five fingers in the order of primates compared with other orders within a more inclusive analysis, or the emergence of knapped lithic industries among the genuses Homo and Australopithecus, or the presence of particular technical knowledge in only some human populations, as may have been the case, for instance, of ploughing techniques, the selection of seeds, sowing and clearing, the domestication and keeping of other animals, etc. during the process of Neolithisation.

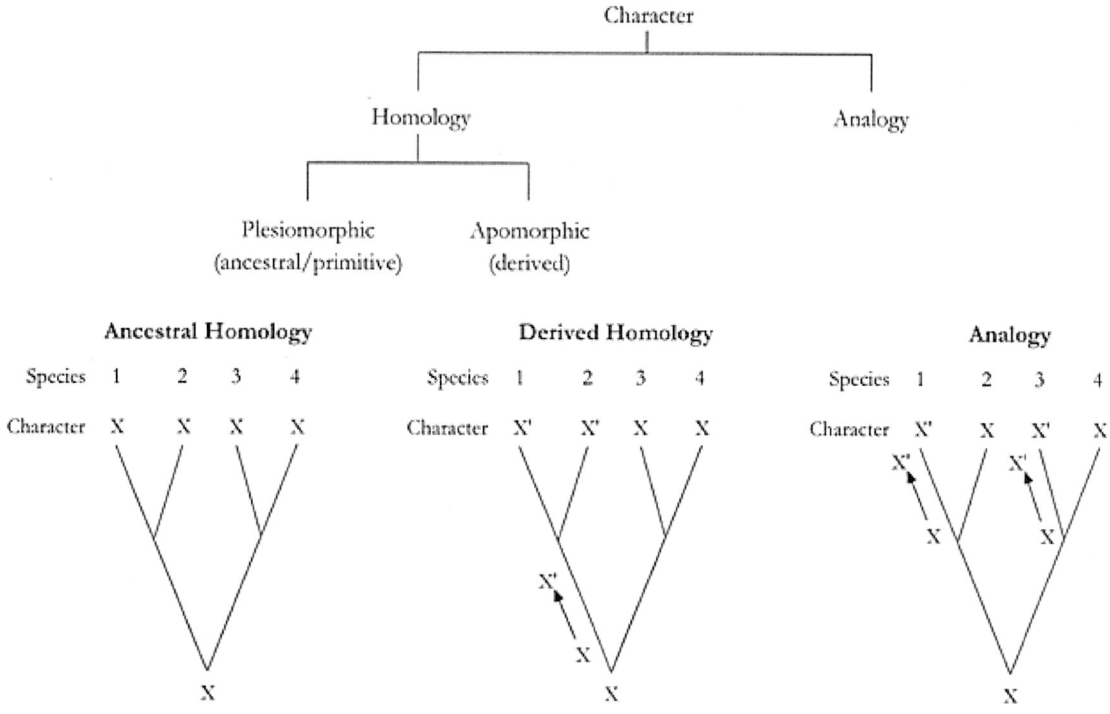

FIGURE 4: CLASSIFICATION OF CHARACTERS IN PHYLOGENETICS, AFTER (O'BRIEN AND LYMAN 2003, FIGURE 3.1.). CLADISTICS, IN PARTICULAR, ONLY CONSIDERS THE APOMORPHIC CHARACTERS, SPECIFICALLY THOSE KNOWN AS SYNAPOMORPHIES (DEFINED IN THE TEXT).

In order to fix a stable criterion (of epistemic value that eliminates ambiguities and subjectivity in the taking of practical decisions), with the intention of eliminating the relativism described above, cladists consider exclusively a single type of apomorphic homologous characters. This is the synapomorphy that gives its name to the character or trait that is shared by all of the sibling taxa of a given group and by its immediate ancestor but not with the second immediate ancestor (this relationship is called symplesiomorphy). In contrast, autapomorphy, a second type of apomorphy, refers to a given character that is present in a taxon but not in its sibling taxa. In cladistics, only synapomorphies are used (in more accessible and known terms these are the shared derived traits or characters), thus constituting what are known as monophyletic groups (Figure 5).

The selection between several possible phylogenetic hypotheses in cladistics is based in the principle of parsimony. Frequently in real case studies, the data provide several possible suggestions of phylogenetic relationships. In other words, a number of different graphical representations or cladograms are obtained, that reflect different hypotheses of how the taxa under study (based on the traits taken into account) relate genealogically to one another. Parsimony is used in the theory and methods of many scientific areas due to its high epistemic value. Faced with the uncertainty of any phenomenon, a simple process that explains the same is favoured over a complex process. In other words, when faced with several different viable explanations for a single process or phenomenon, the simplest or easiest will be selected because it is the most probable. Parsimony is based on the assumption that evolution always operates along the shortest path. This principle is also known as the principle of simplicity and is repeatedly referred to as Ockham's knife (after the 16th century English Franciscan friar William of Ockham) (Hutchinson 1981: 16). Thus, faced with the different cladistic hypotheses, the model that requires the least evolutionary changes will be selected. On the principle of parsimony in phylogenetics see (Panchen 1982; Friday 1982; Felsenstein 1978, 1979, 1983c; Sober 1983, 1988; Maddison, Donogue and Maddison 1984; Farris 1983; Swofford and Maddison 1987; cited in Mayr and Ashlock 1991: 216).

It is also important to mention here Hennig's auxiliary principle, according to which a convergent or parallel evolution must not be assumed a priori; rather in the absence of contrary evidence one must assume homology (Brooks and McLennan 1991: 35).

Cladistics is also the object of several objections. Firstly, and in relation to the above, the epistemic criteria of parsimony is insufficient for some phylogenists of evolutionary taxonomy (these generally base their approaches on maximum likelihood). Basically they argue that not all phenomena can be guided and therefore explained by parsimony; they argue that parsimony is too simple, since it assumes an implicit model that may not necessarily correspond to all of the data (and processes) (e.g. Felsenstein 1982: 388; 1988: 535). However, its defenders underline that this criterion certainly guarantees objectivity in the analyses and decision taking, thus providing a very strong heuristic and methodological instrument (e.g. Farris 1983; 1986; Goloboff 2003), particularly given that the confluence of the principle of parsimony with systematic phylogenetics is compatible with popperian philosophy (Popper 1959; 1963), that is that it possesses the scientific qualities of falsafiability, refutability, and testability (Wiley 1975).

Secondly, because systematic phylogenetics assumes the divergent model of cladogenic evolution, it ignores directly anagenesis or the phyletic evolutionary model. Some criticisms underline this supposed undervaluation of anagenesis. However, even if the existence of an anagenetic evolution were to be accepted as a linear evolution in which a species derives into a single other species, why should we not consider these as the same species? Or rather what makes us think that we are dealing with two species? Moreover, even if this duality is maintained, there is currently no criteria to establish the end and beginning of one and another species (clear cut criteria of separation).

Deriving from this, the cladistic model is criticized for the rigidity of its tree-like structure, which undervalues anagenesis and hybridization. This is related to the problem of homoplasy that can take one of three forms: parallelism, evolutionary convergence and the reversion of states in the characters.

Though often implicit in some of these situations, it is necessary to separate in practical terms a fourth case: the horizontal transmission of information between sibling taxa and coexisting species. This final objection is identified as problematic in the construction of phylogenetic trees, thus implying some methodological flaws such as, for example, the impossibility to estimate descendent taxa by the hybridization of two previous taxa. However, the undervaluation of the processes of anagenesis and hybridation was tackled a long time ago by some cladists through the calculation of Wagner trees and the construction of networks (Kluge and Farris 1969; Farris 1970). Recently, mixed approaches are emerging that use network methods and other graphs of distances (split decomposition), along with phylogenetic trees, in order to solve some of the problems posed by homoplasy, horizontal transmission and more technically polytomies at internodes (Bryant and Moulton 2002; Jobling et al. 2004; Makarenkov and Legendre 2004; Bryant et al. 2005; Huson and Bryant 2006).

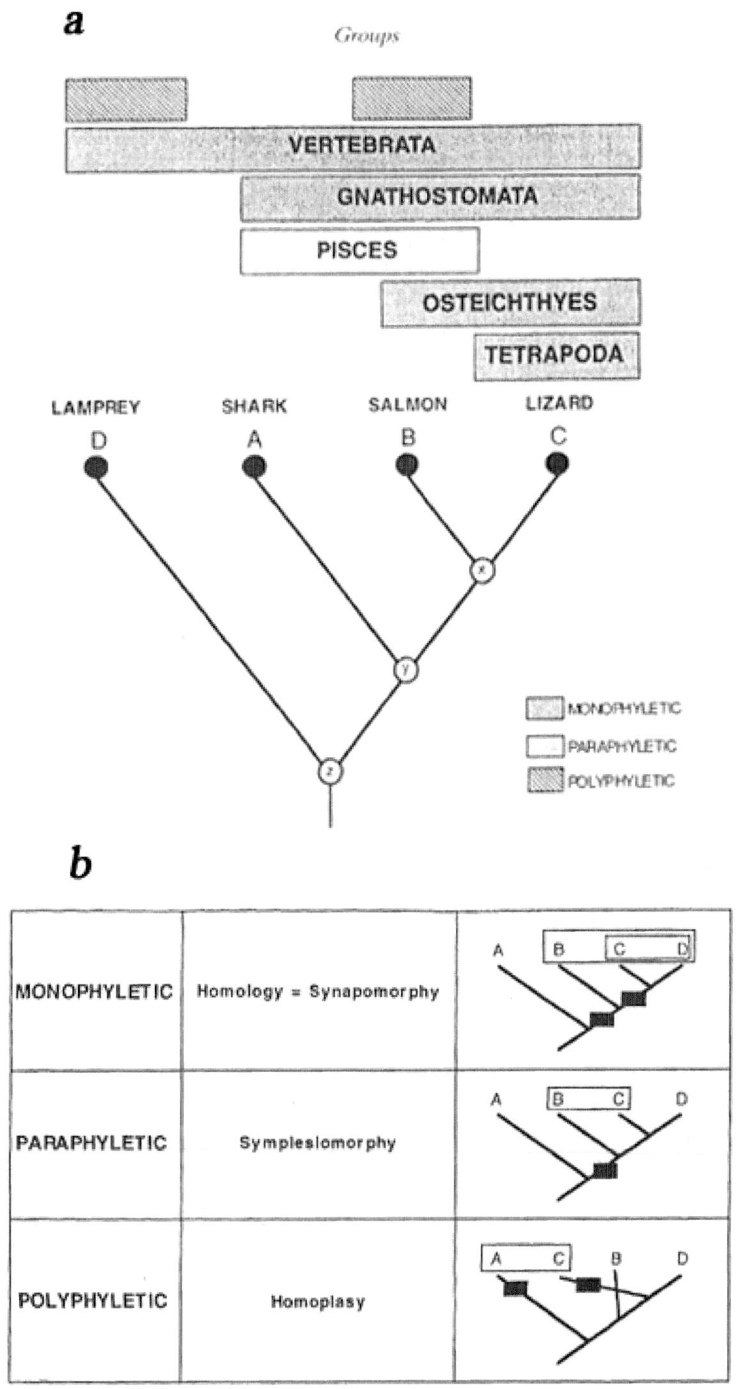

FIGURE 5: (A) TYPES OF TAXONOMICAL GROUPS, AFTER (KITCHING ET AL. 1998, FIGURE 1.8); AND (B) TYPES OF CHARACTERS CONSIDERED IN THE CONSTRUCTION OF EACH OF ONE, AFTER (KITCHING ET AL. 1998, FIGURE 1.10). FIGURE ELABORATED AFTER (KITCHING ET AL. 1998, FIGURES 1.8 AND 1.10.).

Thirdly, the cladistic orthodoxy is often criticized because it considers only monophyletic groups and rejects paraphyletic groups (groups that include symplesiomorphies and/or autapomorphies) (Figure 5). However, the cladists ask why evolutionary taxonomy generally rejects polyphyletic groups and not the paraphyletic groups, when there is no philosophical criteria that serves to differentiate between the two. They conclude that the choice of taxa in paraphyletic groups is arbitrary, subjective and idealistic (Ridley 1996: 388 and ff.).

Finally, some more purely technical criticisms argue over the problem of the long branch attraction, and over the deficiency in the consideration of the different evolutionary rhythms in different phylogenetic branches or lineages. The problem of the long branch attraction is a statistical complication in the algorithms used in the methods of parsimony. Occasionally, when some lineages evolve rapidly, parsimony infers a narrow relationship between them, despite the contradiction with their true evolutionary relationships, thus in sort converting a degree of homoplasy into synapomorphy (Felsenstein 2004; Bergsten 2005).

Evolutionary taxonomy

Evolutionary taxonomy consolidated a few years after systematic phylogenetics, although they developed in parallel. The central figure of this taxonomic tendency was certainly E. Mayr, who in 1969 wrote the first version of Principles of Systematic Zoology. This perspective probably emerged as a critical and alternative response to cladistics, though it also takes into account the evolutionary postulates of change by modification in the descendents. The criticisms were based on the issues that we have dealt with as objections in the previous section. Other criticisms refer to the lack of ranking criteria in the establishing of taxa.

With the intention to eradicate the supposed flaws of cladistics, this new perspective was put forward, based on two fundamental pillars: similarity and phylogeny. Effectively, each of these, respectively, constitutes the principle and exclusive basis of each of the two taxonomical schools presented above. Thus, what Evolutionary taxonomy aims to do is to unite both perspectives under a single approach to classification:

A biological classification is the ordered grouping of organisms according to their similarities and consistent with their inferred descent… It is thus evident that there are two independent sets of criteria responsible for the grouping –similarity and sameness of causation (lines of descent)- on the basis of which organisms can be assembled into taxa. (Mayr and Ashlock 1991: 115).

Evolutionary taxonomists generally carry out their classifications in two steps. First, they establish the similarity between classes, and then these are tested in relation to monophyly, discarding the species that are not formed in this way (Mayr and Ashlock 1991: 130). As mentioned above, Evolutionary taxonomy takes into account the paraphyletic groups (Figure 5) which, we insist, possess homologous characters, both shared derived and ancestral, and in which we may find for example those groups or clades that include some but not all of the descendents from a common ancestor. As we also mentioned above, this approach requires ranking criteria that enable the delimitation of the ranks of the groups. For example, it takes into account the degree of difference between the groups, the size of the gap between them (the taxonomic gap), the evolutionary role in terms of the adaptation of ecosystems, the size of the taxon itself, the stability and the rhythms of evolution, etc. (Mayr and Ashlock 1991: 266).

Evolutionary taxonomy has recently undergone the important statistical development of its methods, thus it could be renamed numerical evolutionary taxonomy. The graphical representations used are known as phylograms. These phylogenetic dendrograms also reflect the length and angle of the internodes, thus expressing for instance the different ratios of evolution. A fundamental point is that evolutionary taxonomy is not restricted to the criteria of parsimony. The increase in the use of statistics in its phylogenetic approaches has led it to use maximum likelihood, thus introducing the theorem of probabilities of Bayes (e.g. Lewis 2001; Holder and Lewis 2003; Ronquist 2004). These new phylogenetic methodologies have become known as comparative phylogenetic methods and are recently being applied to cultural evolution also (Felsenstein 1978; Mace and Pagel 1994; Borgerhoff 2001).

The phylogenetic method with the principle of maximum likelihood requires an explicit model of the evolution of the characters (parameters), thus calculating the likelihood of each state in each position (probability of the data in function of the preestablished model) and selecting, finally, the phylogenetic tree with the highest total score of likelihood. The Bayesian method is based on these principles of maximum likelihood. Effectively, it uses explicit models of evolution and all the information contained in the data. This basic predesigned model constitutes a hypothesis to be tested and later improved. A calculation of the probabilities is made before integrating all of the parameters, and a sample of trees is created by means of the algorithm of the Markov Chains of Monte Carlo (MCMC). This statistically significant selective sample provides the basic properties which can enable the construction of a solution of consensus.

The defenders of this taxonomical school underline some of its advantages. They highlight that its applications use and, moreover, accumulate scientific knowledge, in the sense that the basic approach outlined above may be repeated on the basis of the hypothesis that is being polished. Thus, these taxonomists note that these methods may improve the current models and be applied to other more complex models, with the subsequent comparison between all of them. Moreover, they argue that these methods enable the assessment of the residual uncertainty of the results obtained by parsimony. The consideration of more complex and explicit models of evolution enables the followers of this approach to explore quite a scope of evolutionary processes, some of which were supposedly excluded by the cladistic procedure. For instance, the direction and the order of the evolutionary changes, the ratios of evolution, the correlation between traits (due to processes of coevolution), the estimation of the states of the ancestral characters and the nature itself of the evolutionary process.

However, this taxonomical school also raises a series of objections. First, some researchers directly reject the criteria of similarity for the purpose of classification, even when an attempt is made to align it with the phylogenetic criteria. It must be noted that this perspective is problematic since the two main aspects under study, the model of descendence and the degree of divergence, do not always follow the same models and rhythms (Eldredge and Gould 1972; Gould and Eldredge 1993). It is thus very difficult, if not impossible, to incorporate simultaneously the relative time of the division of the lineage and the rhythm of change since its divergence (Brown and Lomolino 1998; cited in O'Brien and Lyman 2003a: 16). The theoretical entanglement of this approach, as the result of the obliged confluence of the criteria of similarity based on distances and the cladistic criteria based on inheritance, is also often stressed. This combination creates such degree of ambiguity and theoretical and methodological subjectivity (in ad hoc decisions lacking any stable criteria) that for many phylogenetic taxonomists, this approach is directly incorrect, ontologically, and therefore useless in practice. E. Mayr himself acknowledged that there is no proven method to consider simultaneously similarity and genealogy (Mayr and Ashlock 1991: 122). Some other specific objections have centered on the ranking criteria (Wiley 1981: 240-269).

Finally, a further objection concerns, in some way, the general question of the ontological and operational relationship between patterns and processes, specifically the problem of ancestral species. This perspective assumes the statistical search for specific realities of study. The approach departs from the basis that the phenomenon or process (the portion of historical reality under study) may be traced and reconstructed through the continuous perfectioning of models. It must be noted that the assumption of the reliable reconstruction of such a reality has serious consequences. If one aims to reconstruct a evolutionary phenomenon or process exactly as is was (in the phylogenetic sense), one must rely obligatorily on the ancestral species. In contrast to cladistics, which uses patterns to explain processes, this approach cannot exist without real ancestors (O'Brien and Lyman 2005: 105). It is hard to believe that one day we will possess in the paleontological and archaeological records sufficient data on all of the taxa that ever existed.

The key debate of patterns and processess, which took placed in organic evolution several years ago (e.g. Stanley 1979), has recently begun to be addressed in cultural evolution (e.g. Shennan 2009).

Discussion

The Table 1a illustrate the main differences between the three schools of classification in terms of the types of characters used and the construction of groups. Polyphyletic groups are established by phenetics, using indiscriminately all types of characters. And as presented above, monophyletic groups characterise cladistics, while monophyletic and polyphyletic groups are used in evolutionary taxonomy.

The Table 1b compares specifically cladistics and evolutionary taxonomy, after E. Mayr and P.D. Ashlock (1991, Table 10.1). With regards to the classification, it may be appreciated how cladistics use exclusively the existence of monophyly (also called holophyly by other authors), while evolutionary taxonomy also takes into account the number of shared characters. The rank of the taxa in cladistics is categorical, monophyletic groups evidenced by one or more shared derived characters, while it is relative in the other approach. This leads to the consideration by the former of only the processes of bifurcation or division between taxa, while the second also considers the degree of divergence.

Since the bulk of the objections to both schools has been developed at the end of each section, with the inclusion of some references that answer defensively some issues, we shall not insist on these again. Moreover, it is not our intention to confront the different schools, especially cladistics

a

	Groups Recognized			Characters Used		
					Homologies	
Classification	Monophyletic	Paraphyletic	Polyphyletic	Analogies	Ancestral	Derived
Phenetic	Yes	Yes	Yes	Yes	Yes	Yes
Phylogenetic	Yes	No	No	No	No	Yes
Evolutionary	Yes	Yes	No	No	Yes	Yes

b

DIFFERENCES BETWEEN CLADISTIC AND EVOLUTIONARY CLASSIFICATION

	Cladistics	Evolutionary classification
	Analysis	
Aim	Construction of a cladogram	Construction of a phylogram
Parsimony	Important criterion of best cladogram	Testing for homoplasy by various criteria
Major role of synapomorphies	To recognize branching points	To discover and reject convergent polyphyly
	Classification	
Delimitation of taxa	By holophyly	By monophyly and possession of greatest number of shared characters
Minimal number of characters needed	A single synapomorphy	As many characters as possible
Ranking of sister groups	Same categorical rank	Different rank if they differ by sufficient autapomorphies
Kinds of characters used	Only diagnostic synapomorphies	All homologous characters, including ancestral ones
Basis of classification	Branching only	Branching and evolutionary divergence
Stem groups	Always with crown group	Sometimes with sister group
Diagnosis of taxa	Only monothetic admissible	Polythetic also admissible
Autapomorphies	Generally ignored	Important in ranking of taxa
Categorical ranking of taxa	No method available, hence arbitrary	By degree of difference

TABLE 1: (A) DIFFERENCES BETWEEN THE THREE TAXONOMIC SCHOOLS WITH RESPECT TO THE TYPES OF CHARACTERS USED AND THE TYPES OF GROUPS CREATED; AND (B) EXPLICIT COMPARISON BETWEEN CLADISTICS AND EVOLUTIONARY TAXONOMY. FIGURE ELABORATED AFTER (RIDLEY 1996, TABLE 14.1 AND MAYR AND ASHLOCK 1991, TABLE 10.1).

and evolutionary taxonomy. These constitute, in a way, two different perceptions and procedures of study, concerned with different purposes. Thus it is important to mention a very insightful quote in this sense, that we should take on board before facing the taxonomic task:... evolutionary taxonomy and cladistics offer different ways of looking at the world. The former addresses both pattern and process –it tracks "genealogical actors in ecological roles" (Hull 1987: 44)- whereas the latter, strictly speaking, addresses pattern only (process comes from the theory chosen to explain the pattern; typically the theory is Darwinian evolution). (O'Brien and Lyman 2003a: 96).

Thus closing this discussion here, what we shall indicate hereon are some theoretical bibliographic references that may be consulted on the subject. Papers on cladistics are numerous, but the following (as well as the references cited throughout the text) are perhaps the most relevant, not only to find further details on this approach but also to understand the discussion outlined above between the two schools (Wiley 1975; 1981; Platnick 1985; Eldredge and Cracraft 1980; Duncan and Stuessy 1985; Forey et al. 1992; Kitching et al. 1998; Goloboff 2003). Texts on evolutionary taxonomy are also abundant in which the reader may find further critiques of cladistics and thus compare between the two approaches (Ashlock 1979; Charing 1982; Felsenstein 2004; Hull 1979; Mayr and Ashlock 1991; Sosef 1997).

The three philosophies of classification all remain valid today, and the followers of one or the other vary quantitatively depending on the different areas of knowledge and on the consideration of the concept of Darwinian evolution held by each researcher. Each of the taxonomical schools displays different objections and differences with respect to the others. But given that much of this discussion is due mainly to the different philosophical basis of each approach, it is practically impossible to unite them under a global perspective. And effectively to discern which one is the most appropriate for any scientific purpose is almost impossible. It is however important to consider seriously the theory of classification, since it has important implications for the results of our archaeological and anthropological approaches.

Acknowledgements

This paper has been written with the support of the research funding of the Plan Andaluz de Investigación, Consejería de Innovación, Ciencia y Empresa, Junta de Andalucía, Spain.

I wish to thank Michael J. O'Brien and Ruth Taylor for the time and help that they have given me. Both read the manuscript of this paper and made important comments for the clarification of some aspects of the text. Ruth also had the patience to translate the original Spanish text to English. Finally, I would like to thank Hernán Muscio and Marcelo Cardillo for their invitation to participate in this book.

References

Adams, W. Y. 1988: "Archaeological classification: theory versus practice". Antiquity 62: 40-56.

Ashlock, P. D. 1979: "An evolutionary systematist's view of classification". Systematic Zoology 28: 441-450.

Bergsten, J. 2005: "A review of long-branch attraction". Cladistics 21 (2): 163-193.

Borgerhoff Mulder, M. 2001: "Using phylogenetically based Comparative methods in Anthropology: more questions than answers". Evolutionary Anthropology 10: 99-111.

Breuil, H. 1933-1935 : Les peintures rupestres schématiques de la Péninsule Ibérique. Lagny: Fondation Singer-Polignac.

Brooks, D. R., and McLennan, D. A. 1991: Phylogeny, Ecology and behaviour. A research program in comparative Biology. Chicago: University Chicago Press.

Bryant, D., Filimon, F., and Gray, R. D. 2005: "Untangling our past: languages, trees, splits and networks". In R. Mace, C.J. Holden and S. Shennan (Eds.), The evolution of cultural diversity. A phylogenetic approach, pp. 67-84. London: University College London Press.

Bryant, D., Moulton, V. 2002: "NeighborNet: an agglomerative method for the construction of planar phylogenetic networks". In R. Guigó and D. Gusfield (Eds.), Workshop in algorithms for

Bioinformatics (Proceedings of the second international workshop in Rome – Lecture notes in computer science), pp. 375-391. Berlin-New York: Springer.
Bueno, P., Balbín, R., Barroso, R. 2008: "Dioses y antepasados que salen de las piedras". Boletín del Instituto Andaluz del Patrimonio Histórico 67, Especial, pp. 34-45. Sevilla: Junta Andalucía.
Carpenter, E., Schuster, C. 1986: Materials for the Study of Social Symbolism in Ancient and Tribal Art: A Record of Tradition and Continuity. New York: Rock Foundation.
Charing, A.J. 1982: "Systematics in biology: a fundamental comparison of some major schools of thought". In A. Joysey and A. E. Friday (Eds.), Problems of phylogenetic reconstruction, pp. 363-440. London: Academic Press.
Cracraft, J. 1983: "Species concepts and speciation analysis". Current Ornithology 1: 159-187.
— 1987: "Species concepts and the ontology of evolution". Biology and Philosophy 2: 329-346.
Darwin, C. R. 1859: On the origin of species by means of natural selection. London: John Murray.
Dobzhanzky, T. 1937: Genetics and the origin of species. New York: Columbia University Press.
Donoghe, M. J., Kadereit, J. W. 1992: "Walter Zimmerman and the growth of phylogenetic theory". Systematic Zoology 41 (1): 74-85.
Duncan, T., Stuessy, T. F. (Eds.) 1985: Cladistic theory and methodology. New York: Van Nostrand Reinhold Company.
Dunnell, R. C. 1971: Systematics in Prehistory. New York: The Free Press.
Eldredge, N., Cracraft, J. 1980: Phylogenetic patterns and the evolutionary process. New York: Columbia University Press.
Eldredge, N., Gould, S. J. 1972: "Punctuated equilibria: an alternative to phyletic gradualism". In T. J. M. Schopf (Ed.), Models in Paleobiology, pp. 82-115. San Francisco: Freeman, Cooper & Co.
Farris, J. S. 1970: "Methods for computing Wagner trees". Systematic Zoology 19: 83-92.
— 1983: "The logical basis of phylogenetic analysis". In N. Platnick and V. Funk (Eds.), Advances in Cladistics, vol. 2 (Proceedings of the second meeting of the Willi Hennig Society), pp. 7-36. New York: Columbia University Press.
— 1986: "On the boundaries of phylogenetic systematics". Cladistics 2: 14-27.
Felsenstein, J. 1982: "Numerical methods for inferring evolutionary trees". The Quarterly Review of Biology 57: 379-404.
— 1985: "Phylogenies and the Comparative method". American Naturalist 125: 1-15.
— 1988: Phylogenies from molecular sequences: inference and reliability. Annual Review of Genetics 2: 521-565.
— 2004: Inferring phylogenies. Sunderland: Sinauer Associates.
Forey, P. L., Humphries, C. J., Kitching, I. J., Scotland, R. W., Siebert, D. J., and Williams, D. M. 1992: Cladistics: a practical course in systematics (Systematics Association Publication 10). Oxford: Oxford University Press.
Forster, P., Renfrew, C. (Eds.) 2006: Phylogenetic methods and the prehistory of languages. Cambridge: McDonald Institute monographs.
Goloboff, P. A. 2003: "Parsimony, likelihood, and simplicity". Cladistics 19: 91-103.
Gould, S. J., Eldredge, N. L. 1993: "Punctuated equilibrium comes of age". Nature 366: 223-227.
Hennig, W. 1950: Grundzüge einer Theorie der phylogenetischen Systematik.Berlin: Deutscher Zentralverlag.
— 1965: "Phylogenetic Systematics". Annual Review of Entomology 10: 97-116.
— 1966: Phylogenetic Systematics. Illinois: Illinois University Press.
Holder, M., and Lewis, P. O. 2003: "Phylogeny estimation: traditional and Bayesian approaches". Nature Reviews Genetics 4 (4): 275-284.
Hull, D. L. 1979: "The limits of cladism". Systematic Zoology 28: 416-440.
Huson, D. H., and Bryant, D. 2006. "Application of phylogenetic networks in evolutionary studies". Molecular Biology and Evolution 276: 254-267.
Hutchinson, G. E. 1981: Introducción a la ecología de poblaciones. Barcelona: Blume ecología.
Huxley, J. S. 1942: Evolution: The Modern Synthesis. London: Allen and Unwin.
Jobling, M. A., Hurles, M. E., Tyler-Smith, C. 2004: "Making inferences from diversity". In M. A. Jobling, M. E. Hurles and C. Tyler-Smith (Eds.), Human evolutionary genetics: origins, peoples and disease, pp. 170-175. Abingdon: Garland Science, Taylor & Francis Group.
Jones, S., Gray, A. 1994. Clasificación. Madrid: Akal.
Jones, W. 1786: "Third anniversary discourse 'On the Hindus'". Reprinted in The Collected Works of Sir William Jones, 1807, pp. 23-46. London: John Stockdale.

Kitching, I. J., Forey, P. L., Humphries, C. J., Williams, D. J. 1998: Cladistics. The theory and practice of parsimony analysis (Systematics Association Publication 11). Oxford: Oxford University Press.

Kluge, A. G., Farris, J. S. 1969: "Quantitative phyletics and the evolution of Anurans". Systematic Zoology 18: 1-32.

Lamarck, J. B. 1809: Philosophie zoologique. Paris: Dentu.

Lewis, P. O. 2001: "Phylogenetic systematics turns over a new leaf". Trends in Ecology and Evolution 16 (1): 30-37.

Lillios, K. 2004: Esprit (The engraved stone plaque registry and inquiry tool). In: http://research2.its.uiowa.edu/iberian/.

Lipo, C. P., O'Brien, M. J., Collard, M., Shennan, S. J. (Eds.) 2006: Mapping our ancestors: Phylogenetic approaches in Anthropology and Prehistory. New York: Aldine Transaction.

Mace, R., Holden, C. J., and SHENNAN, S. J. 2005 (Eds.): The evolution of cultural diversity: A phylogenetic approach. London: University College London Press.

Mace, R., Pagel, M. 1994: "The Comparative method in Anthropology". Current Anthropology 35 (5): 549-557.

Makarenkov, V., Legendre, P. 2004: "From a phylogenetic tree to a reticulated network". Journal of computational Biology 11 (1): 195-212.

Mayden, R. L. 1997: "A hierarchy of species concepts: the denouement in the saga of the species problem". In M. A. Claridge, H. A. Dawah and M. R. Wilson (Eds.), Species: The units of diversity, pp. 381-424. London: Chapman & Hall.

Mayr, E. 1963: Animal species and evolution. Cambridge: Belknap Press.

— 1969. Principles of systematic Zoology. New York: McGraw-Hill.

Mayr, E., and Ashlock, P. D. 1991: Principles of systematic Zoology (2nd revised edition). Singapore: McGraw Hill International Edition.

Meier, R., and Willman, R. 2000: "The Hennigian species concept". In Q. D. Wheeler and R. Meier (Eds.), Species concepts and phylogenetic theory. A debate, pp. 30-43. New York: Columbia University Press.

Mishler, B. D., and Theriot, E. C. 2000: "The phylogenetic species concept (sensu Mishler and Theriot): Monophyly, Apomorphy, and phylogenetic species concepts". In Q. D. Wheeler and R. Meier (Eds.), Species concepts and phylogenetic theory. A debate, pp. 44-54. New York: Columbia University Press.

O'Brien, M. J., and Lyman, R. L. 2002: "The epistemological nature of archaeological units". Anthropological Theory 2 (1): 37-56.

— 2003a: Cladistics and Archaeology. Salt Lake City: The University of Utah Press.

— 2003b: "Style, function, transmission: an introduction". In M. J. O'Brien and R. L. Lyman (Eds.), Style, Function, Transmission. Evolutionary Archaeological Perspectives (Foundations of Archaeological Inquiry), pp. 1-32. Salk Lake: The University of Utah Press.

— 2005: "Cultural phylogenetic hypotheses in Archaeology: some fundamental issues". In R. Mace, C. J. Holden and S. Shennan (Eds.), The evolution of cultural diversity. A phylogenetic approach, pp. 85-108. London: University College London Press.

O'Brien, M. J., Lyman, R. L., Collard, M., Holden, C. J., Gray, R. D., Shennan, S. 2008: "Transmission, Phylogenetics, and the evolution of cultural diversity". In M. J. O'Brien (Ed.), Cultural transmission and Archaeology. Issues and case studies, pp. 39-58. Washington: Society for American Archaeology.

Popper, K. R. 1963: Conjectures and refutations: the growth of scientific knowledge. London: Routledge & K. Paul.

— 1959: The logic of scientific discovery. London: Hutchinson.

Platnick, N. I. 1985. "Philosophy and the transformation of cladistics revisited". Cladistics 1: 87-94.

Ridley, M. 1996: Evolution. Massachusetts: Blackwell Science.

Ronquist, F. 2004: "Bayesian inference of character evolution". Trends in Ecology and Evolution 19 (9): 475-481.

Schleicher, A. 1863: Die Darwinsche Theorie und die Sprachwissenschaft. Weimar.

Shennan, S. J. 2009. Pattern and process in cultural evolution. (Origins of Human Behavior and Culture 2). Berkeley: University of California Press.

Simpson, G. G. 1951: "The species concept". Evolution 5: 285-298.

— 1961 [1969]: Principles of animal taxonomy. New York: Columbia University Press.

Sokal, R. R., and Sneath, P. H. A. 1963. Principles of numerical taxonomy. San Francisco: W.H. Freeman & Co.

Sosef, M. S. M. 1997: "Hierarchical models, reticulate evolution and the inevitability of paraphyletic supraspecific taxa". Taxon 46 (1): 75-85.

Stanley, S. M. 1979: Macroevolution: pattern and process. San Francisco: W.H. Freeman.

Van Driem, G. 2001: Languages of the Himalayas: An ethnolinguistic handbook of the Greater Himalayan region (2 vols.). Leiden: Brill.

Wheeler, Q. D., and Platnick, N. I. 2000: "The phylogenetic species concept (sensu Wheeler and Platnick)". In Q. D. Wheeler and R. Meier (Eds.), Species concepts and phylogenetic theory. A debate, pp. 55-69. New York: Columbia University Press.

Wiley, E. O. 1975: "Karl R. Popper, systematics, and classification: a reply to Walter Bock and other evolutionary taxonomists". Systematic Zoology 24: 233-243.

— 1981: Phylogenetics: the theory and practice of phylogenetic systematics. New York: John Wiley & Sons.

Wiley, E.O., and Mayden, R. L. 2000: "The evolutionary species concept". In Q. D. Wheeler and R. Meier (Eds.), Species concepts and phylogenetic theory. A debate, pp. 70-89. New York: Columbia University Press.

Yang, Z., and Goldman, N. (Eds.) 2008: "Statistical and computational challenges in molecular phylogenetics and evolution". Philosophical transactions of the Royal Society

ENVIRONMENT, SPACE, HISTORY, AND TECHNOLOGICAL EVOLUTION. THE CASE OF THE PATAGONIAN COAST

Marcelo Cardillo

Consejo Nacional de Investigaciones Científicas y Técnicas (CONICET); Instituto Multidisciplinario de historia y Ciencias Humanas (IMHICIHU). Universidad de Buenos Aires (UBA). e-mail: marcelo.cardillo@gmail.com

Abstract

This work explores the study of evolutionary patterns in technology by means of methodological tools derived from multivariate analysis and cladistic tree reconstruction methods. In this regard, the way in which the change in diversity of stone tools can be explained from the interaction of three dimensions: space, environment, and history, is looked into. Making use of a study case from the Patagonian coast, an explanatory model is generated that allows the role of each of these dimensions to be evaluated.

KEYWORDS: CLADISTICS, MULTIPLE FACTOR ANALYSIS, TECHNOLOGICAL EVOLUTION

Resumen

Este trabajo explora el estudio de patrones evolutivos en la tecnología a través de herramientas metodológicas derivadas de la estadística multivariada y de la cladística. Al respecto, se plantea de qué manera el cambio en la diversidad del instrumental lítico puede explicarse a partir de la interacción de tres dimensiones: espacio, ambiente e historia. Empleando un caso de estudio de la costa patagónica, se genera un modelo explicativo que permite evaluar el rol de cada una de estas dimensiones.

PALABRAS CLAVE: CLADÍSTICA, ANÁLISIS FACTORIAL MÚLTIPLE. EVOLUCIÓN TECNOLÓGICA

Introduction: Environment and technology

Studies on the relationship between the technology of hunter-gatherer groups and the environment go back to Oswalt (1973, 1976). This author proposes that technological complexity (measured according to elements composing an artifact or technounit) shows a relationship with the type of adaptation among human groups. Thus, those groups dependent on mobile resources (and in which the risk of loss is expected to be greater, see Torrence 1989) would possess more complex tools than those that depend on resources with less risk of loss. On a global scale and referring to hunter-gatherer societies such as the author studies, this would explain the greater complexity observed in the hunting tools of human groups depending mainly on marine resources (Oswalt 1976). Torrence (1983, 1989) later related this same phenomenon to the risk of losing resources, showing a non-random relationship on a latitudinal scale, considering the latitudinal increase as a risk proxy, as it involves environments with a lower effective temperature, a shorter growing season and, therefore, greater dependence on mobile resources.

After this (Collard et al. 2005) retested the relationship proposed by Torrence on risk and technology, once more using Oswalt's data, but bringing into play three different hypotheses initially suggested by other researchers as regards residential mobility (Shott 1986), demography (Shennan 2000; v. also Henrich 20004, 2006; Kline and Boyd 2010). The results showed that environmental variability, used as a risk proxy, would be the variable that best explains the technological diversity among the groups that were studied. Still, on exploring this hypothesis again by means of a more limited collection of ethnographic data from the northwestern coast (Collard et al. 2011); they observed no statistically significant relationship. On the basis of these results the authors posit that this is related to the low difference in risk between sectors (due to the spatial scale). This last makes clear the importance of geographical distance in the study of variability.

In Argentine Patagonia, studies carried out on the archaeological record on different spatial scales show that on a macroscale (on a latitudinal level) it is possible to observe variations in the proportion of stone tools assemblages in almost continuous way along the coastal strip. These analyses suggest that space and environment are explanatory factors for the variability of artifacts. In other words, the toolkits tend to diverge from one another as the distance between samples increases, a differentiation that grows if that distance additionally involves environmental variations.

A similar phenomenon has been observed on smaller scales (Charlin et al 2012), between southern Patagonia and the north of Tierra del Fuego. In this particular case, however, the magnitude of the difference between the two sectors is greater than would be explainable by purely spatial or environmental factors (that explains about 10% of stone tool diversity). This suggests other interacting mechanisms, such as the time of separation between human groups on one and the other side of the Magellan Strait.

From a comparative viewpoint, combining phylogenetic and spatial information Cardillo and Charlin (2012) observed indicators of spatial structuration in the morphological variation of middle-late Holocene projectile points. The observed pattern agrees with a model of isolation by distance, since the greater the spatial distance between samples the greater the morphological differentiation (Cardillo & Charlin 2011, 2012). That is, the larger the separation is between groups, the more probable it is to observe a process of divergence, whereas neighboring groups will tend to show parallel trajectories related to the larger information flow between them (Borrowing processes between cultural units). These results show the usefulness of a phylogenetic approximation combined with other statistical tools so as to incorporate the historical dimension in spatial and environmental analyses.

In this work we are putting the usefulness of this comparative methodology in exploring the role of historical, spatial, and environmental dimensions in the variation of artifacts to the test. For this purpose we will attempt to articulate cladistic procedures (historical dimension) and those of multivariate statistics (spatial and environmental variables) using in an exploratory manner late Holocene stone tool samples from the Patagonian coast.

Study area

The Patagonian coast shows important environmental differences, due in part to its enormous extension, and the shape of the continent (Clapperton, 1993). The sector corresponding to the 40°-42° latitude strip is particularly notable for the high biodiversity of marine resources like fish and mollusks (Curtolo, 2004). This supply will presumably have been the same in the past, and isotopic studies carried out on

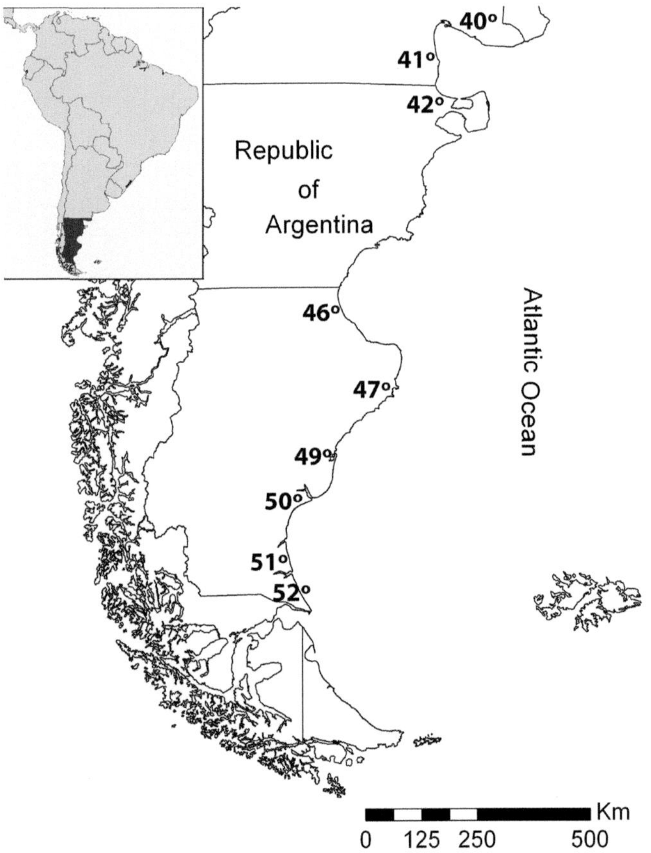

FIGURE 1. LOCATION OF SAMPLES POOLED BY LATITUDE

human remains suggest marine resources made up an important part of human diets (Gómez Otero, 2007; Favier Dubois et al., 2009). The importance of these resources is also seen in the shell deposits between 40° and 49° (Fig. 1), sometimes of great extension and density (Castro et al., 2004; Favier Dubois et al., 2006; Zubimendi, 2010), as well as in the presence of faunal resources such as fish (Gómez Otero, 2007; Scartascini et al., 2009, 2010) and sea-lions (Moreno and Izeta 1999, Gómez Otero, 2007; Moreno, 2008; Borella et al., 2006). In the same way, tools related specifically to the acquisition of coastal resources, such as "skull-crushers" to kill sea-lions (Moreno et al., 2000, 2008) and fish-net weights, have been recovered (Gómez Otero, 1995, 2007, Scartasini and Cardillo, 2009).

As for terrestrial resources, isotopic and archaeofaunal evidence indicates that in the higher latitudes of southern Patagonia the role of mobile resources such as the guanaco was greater, whereas the others have been more complementary (Barberena, 2002, 2008; Borrero et al.,2009). This could be connected, as mentioned, to the environmental characteristics of this area. In this regard, it is important to point out that the southernmost extreme is included in the area of greatest livestock production as, due to the humidity and availability of pastures, it supports an important secondary biomass (PNUD, 85/019 1990, Oliva et al., 2001). For this reason, and if the climatic conditions held out, this region could have supported larger herbivorous populations in the past.

In relation to available chronology, this spans the whole of the Mid-Late Holocene. In the 40° latitude sector datings have been obtained that go from 6000 to 450 years C^{14} B.P. (Favier Dubois et al., 2008). Radiocarbon dates from the Mid-Holocene were also obtained at latitude 47° (Castro and Moreno, 1996-1997; Castro et al., 2007), though the majority belongs to 2000-3000 years C14 B.P. In a like manner, almost all assemblages from latitudes 50°, 51°, and 52° belong to the last 2000 years C14 B.P, with radiocarbon chronologies later than around 1000 years B.P. being far less frequent (Borrero and Franco, 2002, 2007; L´Hereaux and Franco 2002; Mansur et al., 2004, 2007, Franco, 2008; Muñoz, et al 2009, among others). In this work we will take the recent Mid-Late Holocene as a single time-block, assuming there are no significant differences in diversity that might affect samples, considered as time-averaged during this time-span.

The area under study covers some 1300 km of coastline, from the 40° south latitude at the northern shore of the San Matías Gulf, province of Río Negro, down to 52° south latitude, province of Santa Cruz, Argentina. Within this area three great eco-regions can be discriminated according to Bailey (1989). The northern region corresponding to the limit between tropical/subtropical steppe and that included between 41° to 50° southern latitude, which is in the temperate desert. This last sector is on the limit with the temperate steppe. The area between parallels 49°-51° has a climate of the arid Patagonian type, whereas the extreme southern sector (52°) is within the subhumid Magellanic climatic strip. The difference in mean annual temperature between both extremes is some seven degrees less at the southern end of Patagonia (the mean oscilates between 14.2 °C at 40° and 5.4 °C at 52°) (Fig. 1). Of these three eco-regions, the northern sector of North Patagonia (40°) has the greatest plant biomass, including different ligneous and herbaceous species, as well as more abundant annual rainfall (the mean is some 319.3 mm in the north, and 138 mm in the south SMN, PNUD report, 85/019 1990). This primary biomass decreases towards the higher latitudes, although the temperate steppe eco-region has a greater relative humidity than that of the temperate desert (between 41° and 50° latitude). It is therefore expected that the differences observed in temperature, vegetation, and humidity, etc. will have affected human strategies of exploitation in those areas, being a selective factor as was proposed by Rindos (1989).

Materials and Methods

The sample is composed of 1914 tools distributed between nine assemblages used in the statistical analyses. These assemblages pooling the information generated by different researchers all along the Patagonian coast (latitude 40°, 41°, 52° Caracotche et al., 2005; 50° Cardillo, 2009; Franco et al., 2010; 42° Gómez Otero et al., 1999; 46° and 47° Zubimendi, 2010; 49° Lanata et al., 2004; Frank and

Skarbun, 2009; 51° Ercolano and Carballo Marina, 2005). Thus the data used is derived both from personal research as well as data obtained in the mentioned publications.

Each of these assemblages, however, is different in size, the smallest being the sample from parallel 41°, with 26 cases, and the largest that of 47° with 731 tools, the average being 120 tools per assemblage. These samples represent the sum of more than one surface sampling (belonging to high or low density loci, obtained by means of systematic quadrants and transects), but only those sampled systematically, and with the same typological criterion of classification, were selected (i.e., following Aschero, 1975, 1983). Information on type-richness was obtained from tables of absolute frequency, which was then made use of to turn the data into proportions. During the initial handling of obtained information, those publications in which the sampling criteria were not made clear were left aside. A preliminary quantification and descriptive statistics were carried out, from which types present in just a single assemblage or represented in low proportion (less than 1%) were eliminated (see Fig.9 and Table 3 in the appendix).

As we know the simplest way to measure diversity is to count artifact types (richness). This measurement, though informative and easy to interpret, is usually related to the size of the sample (Kintigh, 1989; Hiscock, 2001). Thus, it can be expected that the larger the samples, the greater the probability of observing new classes, especially those of relatively low abundance, as Hiscock points out (2001). In order to compare samples of different size we transformed raw data into proportions for each tool class. Those tools more frequent in each sample will possess a value closer to one, and the less common ones, to 0.

This matrix of proportions was employed in phylogenetic analyses in which each latitude is regarded as a hypothetical taxonomic unit (HTU) described according to the proportion of tools.

Environmental data

With regard to environmental information, four variables registered for each of the samples were used: the mean annual temperature, annual rainfall (from Servicio Meteorológico Nacional database) in addition to measurements derived from these variables such as the growing season and maximum and minimum expected biomass (from Binford, 2001). For the Late Holocene, the palaeoenvironmental evidence suggests that the environment attained analogous values to the present (Mayr et al., 2007), so that the use of these proxies is considered to be suitable for the present purposes. In any case, given the observed correlation in Patagonia between latitude, environmental and ecological variables (Clapperton, 1993) it can be expected that with more detailed palaeoenvironmental evidence, this analysis may be applied to other moments of the Pleistocene-Holocene.

These variables allow a general description of some of the main environmental properties along the Patagonian coast and have been used as predictors. Yet it is to be expected that these environmental variables should be correlated with each other, especially in the case of estimations like biomass, or the duration of the growth period, which are affected by multiple factors, such as temperature and rainfall, as well as a temperature gradient from northern to southern Patagonia. In order to avoid correlation, a PCA was carried out to obtain new orthogonal predictors that might explain the general tendency of environmental change along the space in a non-redundant way (see Fig. 10 and Table 4 in the appendix).

Spatial analysis

It was to be expected that the distance between the sampled sets, the space, should play a central role in the patterns so observed. Sectors closer to each other share similar environmental and geographical conditions, so the spatial component may play an important role in the structuration of archaeological variability. To give an account of spatial variability on different scales and its relationship with the environment, eigenvector-based filtering, or spatial eigenvector mapping (SEVM) (Dormann, et al;

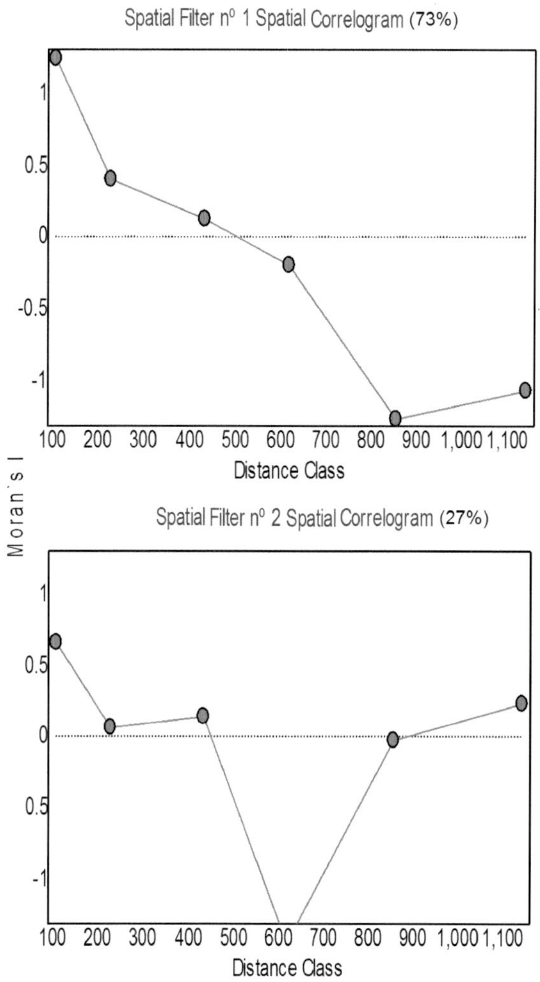

FIGURE 2. TWO FIRST SPATIAL VECTORS OBTAINED FROM SPATIAL COORDINATES

2007) was used. SEVM extracts eigenvectors from distance matrices, created in this particular case from Cartesian coordinates (decimal degrees) obtained for each sample under study. The method implemented here performs a principal coordinate analysis of a connectivity matrix, so that eigenvectors obtained by the analysis express the relationships between samples at different spatial scales, Boccard and Legendre (2002), Diniz-Filho and Bini (2005) and Griffith and Peres-Neto (2006). These eigenvectors can be further used as predictors to take into account the effect of spatial structuration while minimizing autocorrelation in the residuals. Level of autocorrelation is measured by Moran's I spatial autocorrelation index, commonly employed in geography and ecology (Legendre and Legendre, 1998). The SEVM effectiveness to model the species-environment relationship across space in ecology was shown by Dormann et al (2007) and by Diniz-Filho et al (2009) in genetics, among others. SEVM analysis was performed with SAM 4.0 software (Rangel, et al.; 2010).

Cladistic analysis

Phylogenetic analysis has proved to be a powerful tool in the study of evolutionary trajectories of cultural traits (O'Brien and Lyman 2003, O'Brien et al. 2005). The basic supposition behind the application of these methods is that culture is an inheritance system similar to the genetic one observed in biology. In this case copying and the different learning processes enable transgenerational information transfer (Guglielmino et al. 1995, Boyd et al. 1997). Experimental and observational studies indicate that in general terms this system would be sufficiently reliable in maintaining culturally inherited information despite replicative error and horizontal information transmission among individuals and groups (or cultural borrowing, see discussion in Boyd et al. 1997, Muscio 2010).

Therefore, if culture amounts to a more or less coherent inheritance system it is possible to utilize phylogenetic reconstruction to map the trajectories of change, appearance, or extinction of cultural features (Boyd et al 2007). Regarding its application to archaeological data it has been shown that cladistics allows different processes of cultural change to be modeled, such as cultural diversification (Cardillo 2009, Tehrani & Collard 2002, Collard 2006a,b ; Matthews et al. 2011), parallelism, convergence (Collard et al 2006a, b; Temkin & Eldredge 2007; Prentiss 2011, between others) or the role of independent innovation and borrowing in shaping historical patterns to be evaluated (see discussion in Muscio 2010).

For this analysis, each assemblage was taken as an evolutionary unit or HTU (Hypothetical taxonomic unit), while the proportion of tools found in each was used as cladistic characters. We consider that the proportions of stone tools are valid characters, as their frequency in each assemblage represents the average abundance through time of their production and discarding. Preference for one or more artifact classes, their combination within a toolkit, and their change along time are considered evolutionary change (following Rindos 1989). On the other hand, tool classes can be relatively stable features throughout time and space as they possess sufficient integrity to be used as valid characters (see Dunnell 1971 for discussion). This is our hypothesis of primary homology, which will later be tested during cladistic analysis. It can be expected that, if the proportions of tools are characters that carry phylogenetic information, the change thereof will represent shared evolutionary novelties (synapomorphies) among the individuals making up the group. This becomes the empirical proof of the coherence of those, or secondary homology. Those more robust trees are the ones containing the greatest number of synapomorphies. Instead, if these characters vary at random, or being repeated in different groups, they are considered homoplasic. More specifically, homoplasy results from the convergence (independent evolution towards the same combination of characters), reversion or parallelism. Identifying the type of homoplasy in tools also has important information on the evolutionary dynamic and can be modeled by means of different methods (like reticulation analysis, see Gould 2002, Temkin & Eldredge 2007, Prentiss 2011).

The phylogenetic signal in a tree can be measured by different procedures. Two common quantitative measures of character fit are the consistency and retention indices, which measure the quantity of synapomorphies related to the number of character changes, both having a value that varies between 0 and 1. The CI approaches 1 when no reversions or parallelisms of characters take place, whereas the retention index takes into account the number of changes for each character in relation to the total number of steps. If no homoplasy is observed, this index is 1. Also, the bootstrap support was estimated, (n=1000 pseudo-replication) where branches with a support under 50% were collapsed and are represented as polytomies. The character matrix was run via the program TNT (Goloboff et al. 2003).

Later the optimal tree obtained was used to calculate distance matrix (Fig. 3) from the number of nodes between all possible pairs of HTU. The greater the separation between clades, the more nodes there will be between them. This distance matrix was used in an analysis of principal coordinates (PCO) to obtain new orthogonal variables representing the relationships of descent of the HTU in terms of Euclidian distances. Thus the first PCOs represent the general pattern of diversification, the subsequent ones, relations on a smaller scale (Peres-Neto 2006)

Multiple Factorial Analyses

Multifactorial analysis (MFA) is a PCA (in the case of continuous variables) is an analysis of correspondence for qualitative variables (Pagés 2002). This analysis starts out by reducing each of these matrices on its own, retaining the variables that explain a significant portion of the variance, so as to later carry out a global analysis. In each case new weights are assigned to each group of variables (the weight is the first eigenvalue for the PCA of each group) (Pagés 2002). Thus, the maximum axial inertia of each group of variables is equal to 1 (Pagés 2002). As Lê et al (2008, pp 11) observed "…If all the sets of variables are introduced as active elements, without balancing their influence, a single set can contribute quite by itself to the construction of the first axes. In this manner the influence of the groups of variables in the global analysis is balanced and the structure of each group is respected". This analysis is correlative because it excludes any hypothesis of causal influence of a data set on another (Borcard et al. 2011). As in common PCA, the correlation between original variables and new axes can be graphically or numerically represented. Also, the similarity between the geometrical representations derived from each group of variables is measured by the RV coefficient, which varies between 0 and 1. In this case, p-values are obtained by permutations following the procedure suggested by Borcard and co-authors (2011)

Multivariate multiple regression and hypothesis testing

Finally an explanatory model of phylogenetic variance was proposed, making use of all spatial and environmental variables employed in multifactorial analysis (the three components of phylogenetic variance, and the two spatial and environmental ones). Space and environment were treated as independent variables, whereas the three phylogenetic vectors were the dependent ones. In order to adjust the multiple regression model, canonical redundancy analysis was used (Legendre & Legendre 1998). This method, by means of multiple regression, seeks to correlate as closely as possible the main components obtained from the original matrix with the independent variables. In this way the aim is to maximize the variance explained by the latter (in our case, the latitude) from the linear combination of dependent variables (phylogenetic vectors), (see Legendre & Legendre 1998). A conditional step by step regression model was used. In this case the optimal model was the one that would maximize the explained variance (measured by means of R^2 statistics), with the lowest possible p value. The null hypothesis for the regression model affirms that neither environment nor space explain a significant proportion of the branching pattern described by the phylogenetic tree. The Goodness of fit was estimated by means of ordinary R^2 and adjusted R^2. The latter takes into account the number of degrees of freedom and is preferable to simple R^2 in the context of multivariate multiple regression (see Borcard et al. 2011)

Statistical analyses were done with the program R. version 2.15.0 (R developed core team 2005).

Analysis

Results of the cladistic analysis (Fig. 4, 5) suggest a low phylogenetic signal, the values of the consistence index being 0.59 and that of retention 0.50. This consistency index shows somewhat less than half the changes (41%) in the proportion of tools are homoplasic. Values of RI are similar and point to a relatively high frequency (double the optimal) of steps in the evolution of the characters.

These low values coincide with the average bootstrap support obtained (32%) in which all branches with less than 50% support are shown as collapsed. Results indicate a relatively high proportion of homoplasy, whereas the observed geographical order suggests that the branching pattern can be explained by geographical or environmental factors (see below).

The most parsimonious tree (only one tree was obtained) was then used to obtain distances between terminal HTUs (tips). These distances were later employed in a PCO to get orthogonal numeric vectors that would describe the principal variation axes of the tree. The three first dimensions of PCO explain 86% of the total variation and capture the general grouping pattern observed in the tree (Fig. 6).

Multifactorial analysis, which permits the relationship among the three groups of analysed variables to be considered, shows that the first two axes explain 84% of the total variance in the proportion of artifacts (Fig. 7, 8). In the figure (7) it can be seen that the positive values of the first dimension are explained mainly by the spatial correlation and the environmental variables (which would affect middle and low latitudes to a greater degree) whereas high latitudes would be better explained by the first two phylogenetic vectors (Fig. 7).

On observing the contribution of each variable to the total variance (Table 1), it can be seen that the greatest variation is explained by the first spatial filter (34.45%), the third phylogenetic one (26.80%) and the first environmental (34.22%).

In the same way, these three groups of variables are significantly correlated (Table 2), which suggests that the descent and diversification pattern observed in the phylogeny is statistically related with the spatial distance and the environmental variables.

Finally a multivariate regression model was generated in order to explain the relationship of the phylogenetic vectors with the environment and space. The optimal model is the one combining EF1 and the second Pcamb2 ($F=4.461$, $p=0.01$) with an explained variance of R^2 0.60 (60%) and adjusted

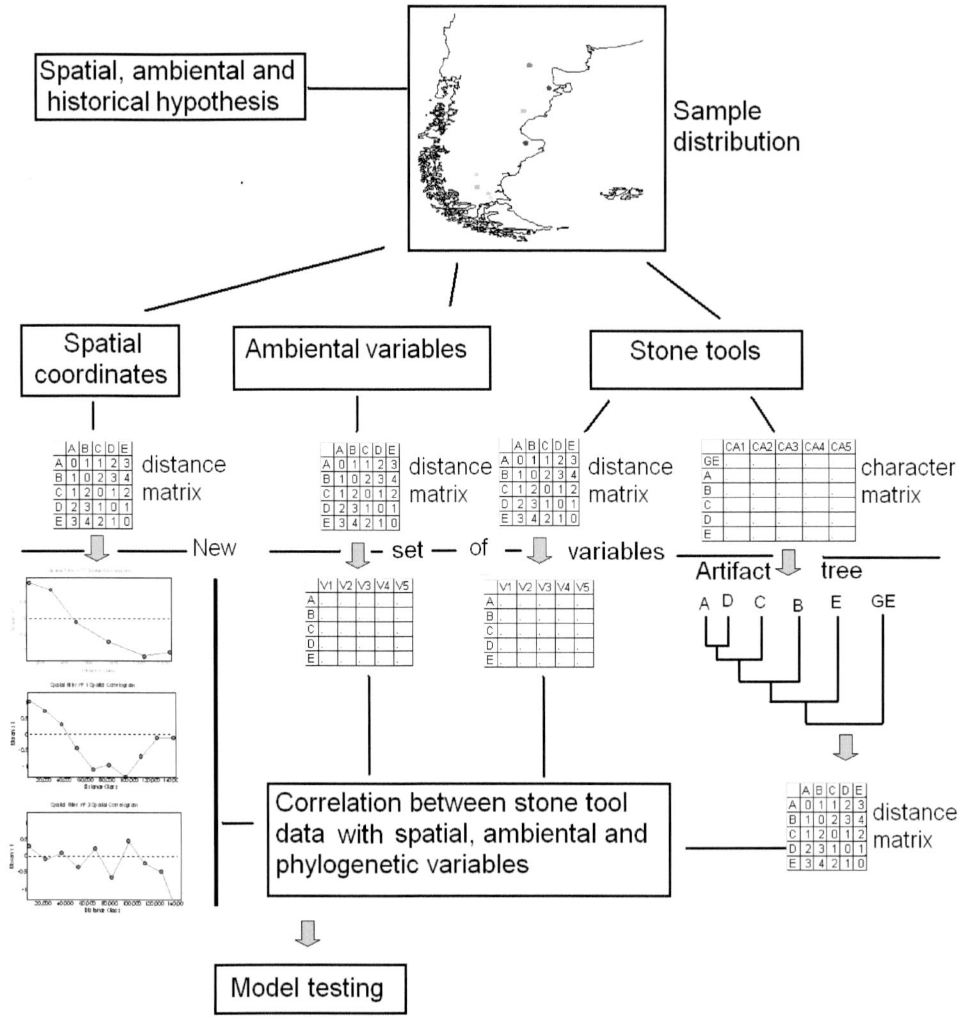

FIGURE 3 THREE-STAGE METHODOLOGICAL SCHEME. GATHERING DATA, GENERATING SPATIAL END ENVIRONMENTAL CORRELATION MATRICES, AND PHYLOGENETIC DISTANCE

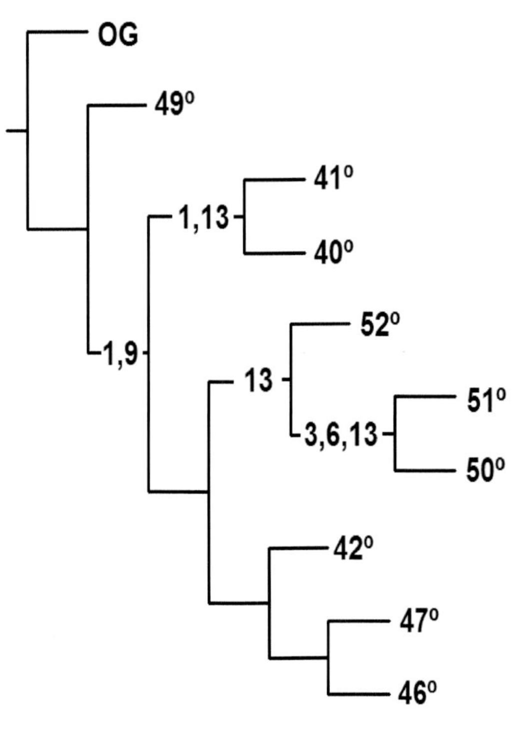

R^2 of 0.45%. Of this, 28% is explained by the spatial distance and the remaining 21% by the environment.

Discussion and conclusions

The combined use of spatial, environmental, and phylogenetic vectors make account for the observed variability in the proportion of stone tools along the latitudinal axis. On the one hand the three measurements of tree support suggest a considerable level of homoplasy in the characters, the exception being the most separated latitudinal groups that conforms well supported clades.

FIGURE 4. A MORE PARSIMONIOUS TREE. SYNAPOMORPHIES OF THE NODES ARE MAPPED WITH BOOTSTRAP SUPPORT ABOVE 50%. 1 DENTICULATE, 9 RABOT, 13 SIDE SCRAPER, 3 RETOUCHED FLAKES, 6 ANVIL.

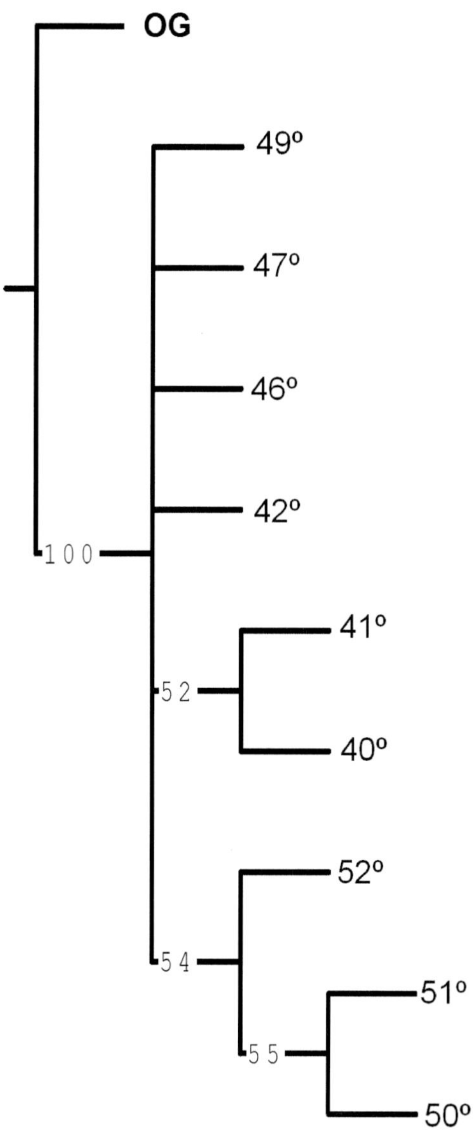

Figure 5. Bootstrap tree. Branches with less than 50% have been collapsed

	Dim.1	Dim.2	Dim.3	Dim.4	Dim.5
EF1	34.45	0.44	0.08	47.65	10.45
EF2	0.56	56.12	4.52	10.40	17.57
PCO1	1.21	0.12	57.93	3.84	2.15
PCO2	2.57	20.08	0.20	4.70	57.75
PCO3	26.80	8.66	24.12	17.62	8.45
PC1	34.22	1.98	13.10	7.92	1.06
PC2	0.19	12.61	0.04	7.87	2.58
%	**100**	**100**	**100**	**100**	**100**

Table 1. Relative contribution to each variable to first five dimensions of multifactorial analysis. EF spatial filters, PCO Principal Coordinates of distance matrix between nodes. PC Principal Components of environmental variables

RV values	Space	PhyVs	AmbV
Space		0.004	0.003
PhyVs	0.61		0.006
AmbV	0.64	0.62	1

Table 2. Correlation between the three groups of variables. PhyVs phylogenetic variables summarized in the first three PCO axes, AmbV Ambiental variables of the two first Principal Component Analysis

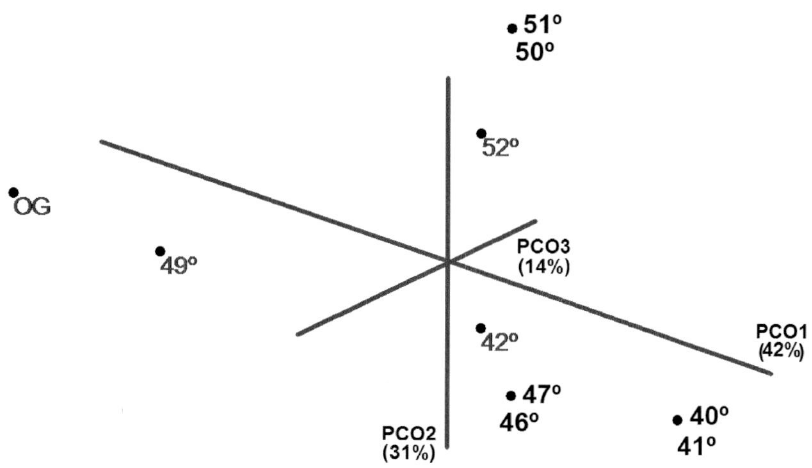

Figure 6. Three first main coordinates selected for factorial analysis (87%) derived from the node-to-node distance matrix

The level of homoplasy suggests on the one hand that the proportion of tools do not carry a good phylogenetic signal (see discussion in Barrientos, in this volume). We believe that the correlation observed between phylogenetic vectors, and environmental and spatial variables suggest that the pattern of grouping is structured by the environment as much as by the spatial dimension considered. The results of multiple regression analysis support this hypothesis, and suggesting that at least the 45% of the phylogenetic variability is explained by the combination of the broad spatial scale (latitudinal gradient of the first spatial filter) and the environmental variability of the second PCamb.

If the synapomorphies of the better resolved sections of the tree are taken into into account, it is possible to consider that the greater proportion of side-scrapers is an inherited average in the clades of the higher latitudes and denticulates of the lower ones. Both would be aspects that

evolved independently in these two areas, a pattern that would be indistinguishable on smaller scales. Also horizontal transmission related to spatial distance may explain paralellism in medium latitudes. Regarding the environment the second PCamb basically represented the differences of the Growing season and biomass between latitudes. In this sense, the stone tools proportion can have evolved as part of human adaptations to spaces with a low biomass and a short growing season (in high latitudes) and a greater biomass and lengthier growing season (in lower latitudes)

If the functional dimension of tools is the most important one, it is understandable that homoplasy in intermediate latitudes should be related to the fact that environmental differences in neighboring spaces are not sufficiently marked to show strong differences in local adaptations.

In consequence it is possible that the lack of resolution observed in intermediate latitudes (42° to 49°) can be related to the little spatial and environmental distance between assemblages. This last coincides with prior analyses carried out on these data (Cardillo 2011), which suggest the preponderance of a spatio-environmental gradient in the proportion of stone tools along the Patagonian coastline, more than breaks or discontinuities. Likewise, this pattern is congruent with what has been observed in projectile points in which the differentiation between high and low/medium latitudes of the coast is what possesses the greatest resolution, while the variability of intermediate latitudes mostly to be explained by spatial scale factors.

It is possible the results indicate that, though adaptive factors and horizontal transmission have been all-important, there exists a tendency towards divergence in technology on the broader spatial scale which is to be found in the proportion of certain types of tools.

This might be explored with characters that contain potentially greater phylogenetic information as could be expected with decorative features, and technical attributes related to production and resharpening of tools (see Barrientos, in this volume), rock art, etc.

Conclusions

The results obtained suggest that the variability of artifacts can be explained by the interaction of environmental, spatial, or phylogenetic dimensions. Results indicate space and environment played an important role in the observed pattern of technological differentiation, at least on a broader scale. The incorporation of new lines of evidence as well as the building of more robust phylogenetic models will serve to contrast this hypothesis.

Acknowledgements

This research was supported by CONICET.

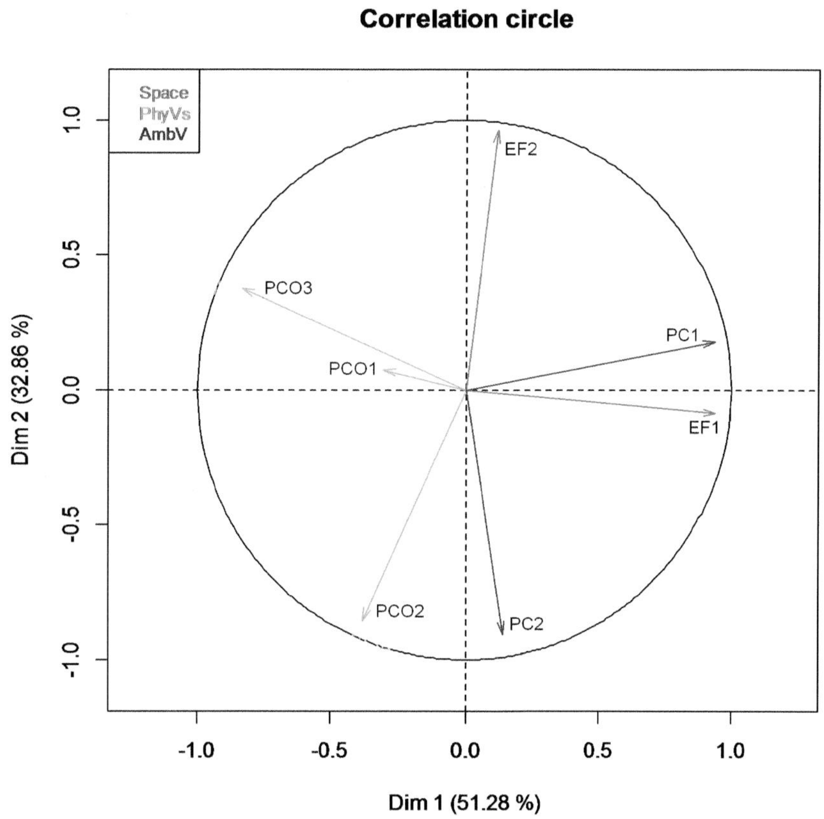

FIGURE 7. CORRELATION CIRCLE BETWEEN THE VARIABLES AND THE FIRST TWO MFA DIMENSIONS. EF SPATIAL FILTERS, PCO PRINCIPAL COORDINATES OF DISTANCE MATRIX BETWEEN NODES. PC PRINCIPAL COMPONENTS OF ENVIRONMENTAL VARIABLES.

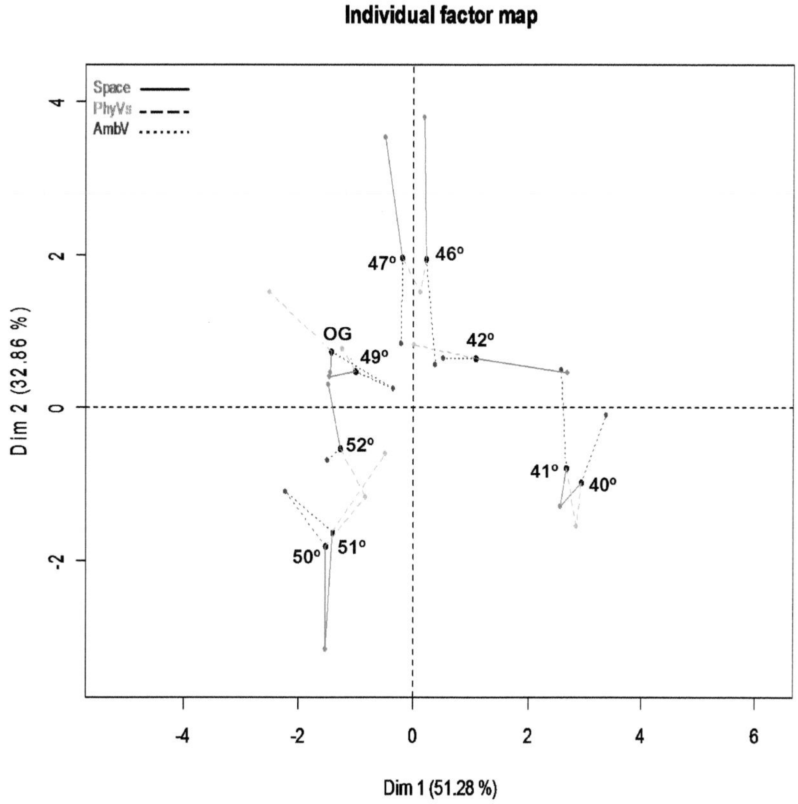

FIGURE 8. PARTIAL INDIVIDUALS RELATED TO THE FIRST TWO COORDINATES OF THE MULTIFACTORIAL ANALYSIS. THE LENGTH OF THE LINES INDICATES THE INFLUENCE OF EACH GROUP OF VARIABLES IN THE LOCATION OF THE CASES.

Appendix

Tool type	40°		41°		42°		46°		47°		49°		50°		51°		52°		Σ
	f	%	*f*	%	*f*	%	*f*	%	*f*	%	*f*	%	*f*	%	*f*	%	*f*	%	*f*
Denticulate	23	13.9	3	11.5	1	0.8	3	2.7	34	4.7	0	0	9	3.5	0	0	10	3	84
Notch	25	15.2	0	0	0	0	3	2.7	10	1.4	0	0	4	1.6	0	0	1	0.3	43
Retouched Tools	65	39.4	9	34.6	24	20	10	8.8	98	13.4	18	34	37	14.5	0	0	79	23.8	340
Front Scraper	13	7.9	6	23.1	24	20	43	38.1	322	44	24	45.3	14	5.5	28	23.7	59	17.8	533
Unifacial Core-Flake tool	0	0	0	0	1	0.8	0	0	0	0	0	0	3	1.2	0	0	1	0.3	5
Anvil	0	0	0	0	4	3.3	0	0	0	0	0	0	3	1.2	3	2.5	1	0.3	11
Biface	0	0	0	0	10	8.3	2	1.8	46	6.3	0	0	12	4.7	0	0	1	0.3	72
Bola	1	0.6	0	0	3	2.5	1	0.9	0	0	0	0	1	0.4	2	1.7	5	1.5	13
Rabot	2	1.2	2	7.7	1	0.8	1	0.9	4	0.5	0	0	5	2	1	0.8	2	0.6	18
Preform	14	8.5	1	3.8	0	0	4	3.5	3	0.4	0	0	7	2.7	14	11.9	17	5.1	60
Burin	7	4.2	0	0	20	16.7	2	1.8	14	1.9	0	0	7	2.7	0	0	7	2.1	57
Chopper	1	0.6	0	0	0	0	0	0	8	1.1	0	0	4	1.6	0	0	3	0.9	16
Side Scraper	10	6.1	4	15.4	9	7.5	37	32.7	155	21.2	10	18.9	131	51.4	67	56.8	139	41.9	561
Hammer	4	2.4	0	0	12	10	1	0.9	3	0.4	1	1.9	7	2.7	0	0	2	0.6	30
Projectile Point	0	0	1	3.8	11	9.2	6	5.3	34	4.7	0	0	11	4.3	3	2.5	5	1.5	70
Σ	165	100	26	100	120	100	114	100	731	100	53	100	255	100	118	100	332	100	1914

TABLE 3. TOOL FREQUENCY AND TOOL PROPORTION BY LATITUDE

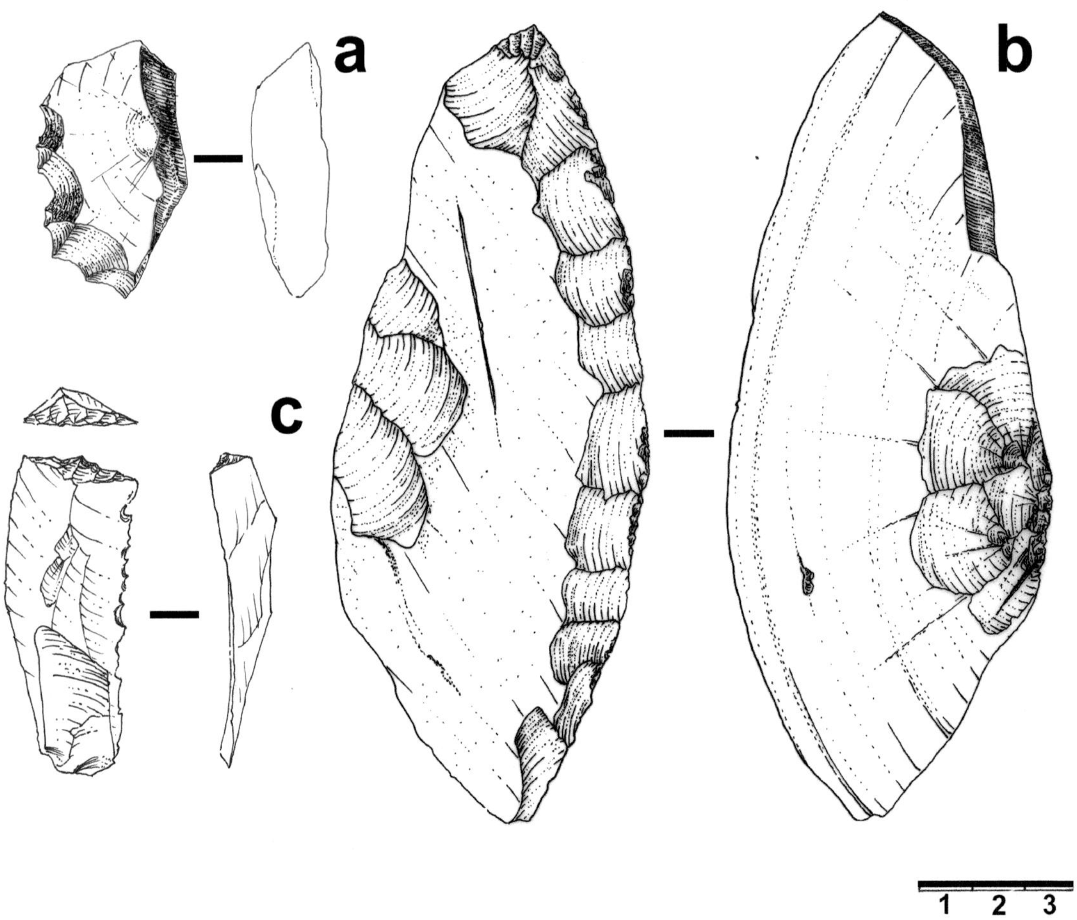

FIG 9: MOST COMMON TOOLS IN SAMPLE A) TYPICAL DENTICULATE. B) TYPICAL FRONT-SCRAPER, C) BIG SIDE-SCRAPER WITH RETOUCHED BULB AREA TO ENHANCE HANDHELD MANIPULATION, MOST COMMON IN HIGHER LATITUDES OF THE STUDIED AREA

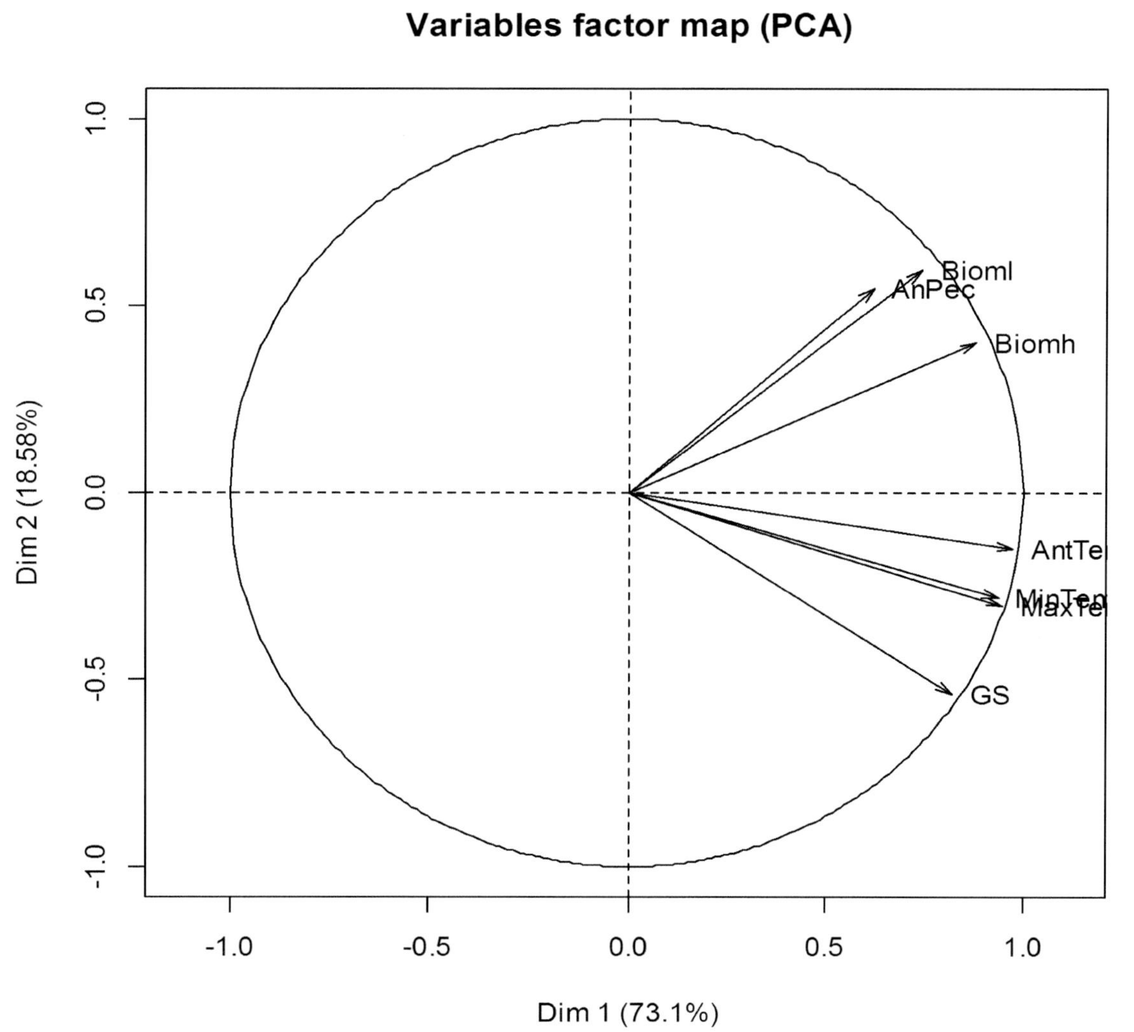

FIGURE 9. FIRST TWO DIMENSIONS PC ANALYSIS RESULTS WITH ENVIRONMENTAL VARIABLES. ANTTEMP ANNUAL MEAN TEMPERATURE, MAX TEM ANNUAL MEAN MAXIMUN TEMPERATURE, MIN TEMP ANNUAL MEAN MINIMUM TEMPERATURE, ANPEC ANNUAL MEAN ANNUAL PRICIPITATION, GS GROWING SEASON, BIOML LOWEST EXPECTED BIOMASS, BIOMH HIGHEST EXPECTED BIOMASS

	Axis 1	Axis 2
AntTemp	0.9755	-0.149
MaxTem	0.949	-0.3028
MinTemp	0.9369	-0.2801
AnPec	0.6206	0.5479
GS	0.8202	-0.5395
Bioml	0.7436	0.5959
Biomh	0.8812	0.4025

TABLE 4. CORRELATION BETWEEN FIRST TWO PC AXIS AND ENVIRONMENTAL VARIABLES. ANTTEMP ANNUAL MEAND TEMPERATURE, MAX TEM ANNUAL MEAN MAXIMUN TEMPERATURE, MIN TEMP ANNUAL MEAN MINIMUM TEMPERATURE , ANPEC ANNUAL MEAN ANNUAL PRICIPITATION, GS GROWING SEASON, BIOML LOWEST EXPECTED BIOMASS, BIOMH HIGHEST EXPECTED BIOMASS.

References

Binford, L. R. 2001. Constructing frames of reference: analytical method for theory building using ethnographic and environmental data sets. Berkeley, CA: University of California Press

Borcard, D., P. Legendre, C. Avois-Jacquet and H. Tuomisto. 2004. Dissecting the spatial structure of ecological data at multiple scales. Ecology 85 (7): 1826-1832.

Borcard, D; Gillet, F and Legendre, P. 2011. Numerical ecology with R. Springer.

R. Boyd, P.J. Richerson, M. Borgerhoff-Mulder, and W. H. Durham. .1997. Are Cultural Phylogenies Possible? In: Human by Nature, Between Biology and the Social Sciences, P. Weingart, P.J. Richerson, S.D.Mitchell, and S. Maasen, eds. Lawrence Erlbaum Associates: Mahwah, NJ, pp. 355–38

Cardillo, M. 2009. Variabilidad en la manufactura y diseño de artefactos en el área costera patagónica. Un enfoque integrador. Unpublished doctoral dissertation. Facultad de Filosofía y Letras, Universidad de Buenos Aires.

Cardillo, M. 2011. Exploring latitudinal patterns of lithic technology variation in continental coastal Patagonia, Argentina. Journal of Archaeological Science 38: 2675-2682.

Cardillo, M. and J. Charlin. 2010. Diversificación morfológica en las puntas de proyectil de Patagonia. Poster presented at Primer Encuentro de Morfometría "Morfometría geométrica y estudios filogenéticos", La Plata.

Cardillo, M. and J. Charlin. 2011. Morphological diversification of stemmed projectile points of Patagonia: Assessing temporal and spatial patters by means of phylogenies and comparative methods. Paper presented at 1st workshop "Multidisciplinary approaches to the study of stone age weaponry", organized by R. Lovita and K. Sano, Institut für Vor- und Frühgeschichte, University of Mainz, Mainz, Germany.

Charlin, J., K. Borrazzo and M. Cardillo. 2012. Exploring size and shape variations in Late Holocene projectile points in Northern and Southern coasts of Magellan Strait (South America). Paper presented at XVI World Congress UISPP. Florianopolis, Brazil.

Clapperton, C.M. 1992 La última glaciación y deglaciación en el estrecho de Magallanes: implicaciones para el poblamiento de Tierra del Fuego. Anales del Instituto de la Patagonia 21: 113-128.

Collard, M., Kemery, M. D. and S. Banks. 2005. Causes of toolkit variation among hunter–gatherers: a test of four competing hypotheses. Canadian Journal of Archaeology. 29, 1–19.

Collard M., Buchanan B, Morin J and A. Costopolous. 2011. What drives the evolution of hunter–gatherer subsistence technology? A reanalysis of the risk hypothesis with data from Early Contact era populations in the Pacific Northwest.

Collard, M., S.J. Shennan & J. Tehrani. 2005. Branching versus blending in macroscale cultural evolution: a comparative study. In C.P Lipo, M.J. O'Brien, M. Collard & S.J. Shennan (Eds.), Mapping Our Ancestors: Phylogenetic Methods in Anthropology and Prehistory (pp. 53-65). New York: Aldine Transaction

Collard, M. 2006. Branching, blending, and the evolution of cultural similarities and differences among human populations. Evolution and Human Behavior, 27 (3), 169-184.

Diniz-Filho, J.A.F. and L.M. Bini. 2005. Modelling geographical patterns in species richness using eigenvector-based spatial filters. Global Ecology and Biogeography 14: 177-185.

Goloboff, P.A., Farris, J.S. &. Nixon, K. C. 2008. TNT, a free program for phylogenetic analysis. Cladistics, 24, 774–786

Goloboff P A, Camilo I. Mattoni and A S Quinteros. 2006. Continuous characters analyzed as such Cladistics.V 22, 6, 589–601

Griffith, D.A., & Peres-Neto, P. 2006. Spatial modeling in Ecology: the flexibility of eigenfunction spatial analyses. Ecology, 87, 2603-2613.

Guglielmino, C.R.; Viganotti, C.; Hewlett, B.; Cavalli-Sforza, L.L. 1995. Cultural variation in Africa: Role of mechanism of transmission and adaptation. Proc. Natl. Acad. Sci., 92, 7585-7589

Henrich, J. 2004. Demography and Cultural Evolution: Why adaptive cultural processes produced maladaptive losses in Tasmania. American Antiquity, 69 (2): 197-21.

Henrich, J. 2006. Understanding cultural evolutionary models: a reply to Read's critique. Am. Antiquity 71,771–782.

Harmon, M. J., VanPool, T. L., Leonard, R. D., VanPool, C. S. & Salter, L. A. 2006. Reconstructing the flow of information across time and space: a phylogenetic analysis of ceramic traditions from prehispanic western and northern Mexico and the American southwest. In C. P. Lipo, M. J. O'Brien, S. Shennan & M. Collard (Eds.), Mapping our ancestors: phylogenetic approaches in anthropology and prehistory (pp. 209-229). New Brunswick: Aldine Transaction.

Collard,M., Shennan,S.J., Tehrani,J.J. 2006a. Branching, blending, and the evolution of cultural similarities and differences among human populations. Evolution and Human Behavior 27, 169-184

Collard,M., Shennan,S.J., Tehrani,J.J. (2006b). Branching versus blending in macroscale cultural evolution: a comparative study. in Lipo,C.P.,O'Brien,M.J., Collard,M., Shennan,S.J. (ed.) Mapping our Ancestors: Phylogenetic Methods in Anthropology and Prehistory. New Brunswick, NJ: Aldine Transaction, 53-64

Goloboff, P., S. Farris, and K. Nixon. 2003. TNT (Tree analysis using New Technology).Ver 1.1. http://www.zmuc.dk/public/phylogeny/tnt/

Gould S.J. 2002. The Structure of Evolutionary Theory. Cambridge MA: Harvard University. Press.

Legendre, P. and L. Legendre. 1998 Numerical Ecology. 2nd English Edition. Elsevier

Jordan P. & S.J. Shennan. 2009. Diversity in Hunter-Gatherer Technological Traditions: Mapping Trajectories of Cultural 'Descent with Modification' in Northeast California. Journal of Anthropological Archaeology, 28 (3), 342-365.

Kline, M. & Boyd, R. 2010. Population size predicts technological complexity in oceania, Proceedings of the Royal Society (B), 277, 2559–2564.

Lê, S.L; Josse, J; Husson, F. 2008. FactoMineR: an R package for multivariate analysis. Journal of Statistical Software. V 25(1), pp 1-18. http://www.jstatsoft.org/

Lycett, S.J. 2009. Understanding ancient hominin dispersals using artefactual data: a phylogeographic analysis of Acheulean handaxes. PLoS ONE, 4 (10), 1-6.

Meghin, O.F A. 1952. Fundamentos cronológicos de la prehistoria de Patagonia. Runa, 5, 23-43.

Matthews L. J.; Tehrani J.J;, Jordan F.M; Collard M & Nunn Ch. L. 2011. Testing for Divergent Transmission Histories amongCultural Characters: A Study Using BayesianPhylogenetic Methods and Iranian Tribal Textile Data. PLoS ONE. V6(4), pp 1-10.

Muscio, H. 2010. Transferencia Horizontal, Cladismo y Filogenias Culturales. In Clasificación y Arqueología: Enfoques y Métodos Taxonómicos a la Luz de la Evolución Darwiniana. Capitulo 9 GarcíaRivero, D. y Escacena Carrasco, J.L. (Eds). Prensa de la Universidad de Sevilla, Sevilla, pp 223-251.

Pagés J. 2002. Analyse factorielle multiple appliquée aux variables qualitativeset aux données mixtes. Rev Statistique appliquée, L (4), 5-37.

O'Brien, M. & Lyman, R. 2003. Cladistics and Archaeology. Salt Lake City: University of Utah Press.

O'Brien, M., Collard, M. & S.J. Shennan. 2005. Mapping Our Ancestors: Phylogenetic Methods in Anthropology and Prehistory. New York: Aldine Transaction.

Orquera, L.A. 1987. Advances in the Archaeology of the Pampas and Patagonia. Journal of World Prehistory, 1, 333-413.

Oswalt, W. H. 1976 An anthropological analysis of food getting technology. New York, NY: Wiley.

Oswalt, W. H. 1973 Habitat and technology: the evolution of hunting. New York, NY: Holt, Rinehart, and Winston.

Pallo, M. C. 2011 Modelo actualístico de limitantes invernales. Implicancias en el uso humano del espacio durante el Holoceno tardío en Patagonia meridional. Poster presented in the VIII Jornadas de Arqueología de la Patagonia, Malargüe, Argentina.

Pallo, C. and J. Charlin. 2010. Distribución de artefactos y variabilidad ambiental en el extremo sur de Patagonia (Pali Aike, Santa Cruz). In: J. R. Bárcena and H. Chiavazza (eds.), Arqueología Argentina en el Bicentenario de la Revolución de Mayo, tomo I, cap. 1, pp. 31-36. Facultad de Filosofía y Letras de la Universidad Nacional de Cuyo, INCIHUSA-CONICET y ANPCyT, Mendoza.

Peres-Neto, P.R. 2006. A unified strategy for estimating and controlling spatial, temporal and phylogenetic autocorrelation in ecological models. Oecologia Brasiliensis 10: 105-119

Pérez, S.I., & Monteiro, L.M. 2009. Nonrandom factors in modern human morphological diversification: a study of craniofacial variation in southern South American populations. Evolution, 63, 978–993.

Rangel, T.F., J.A.F. Diniz-Filho and L.M. Bini. 2010 SAM: a comprehensive application for spatial analysis in macroecology. Ecography 33: 46-50.

R Development Core Team., 2005. R: A language and environment for statistical computing, reference index version 2.2.1. URL. R Foundation for Statistical Computing, Vienna, Austria, ISBN 3-900051-07-0. http://www.R-project.org.

Read, D. 2008. An interaction model for resource implement complexity based on risk and number of annual moves. American Antiquity 73, 599–625.

Rangel, T.F., Diniz-Filho, J.A.F., & Bini, L.M.. 2010. SAM: a comprehensive application for spatial analysis in macroecology. Ecography, 33, 46-50.

Shott, M. 1986. Technological organization and settlement mobility: an ethnographic examination. Journal of Anthropological Research 42, 15–51.

Rindos, D. 1989. Diversity, variation and selection. In R.D Leonard & G.T Jones (Eds.), Quantifying Diversity in Archaeology. Cambridge: Cambridge University Press.

Shennan, S.J. 2000. Population, culture history and the dynamics of culture change. Current Anthropology, 41, 811-835

Tehrani J, Collard M. 2002. Investigating cultural evolution through biological phylogenetic analyses of Turkmen textiles. J Anthropol Archaeol 21: 443–463.

Temkin I; and Eldredge, N. 2007. Phylogenetics and Material Cultural Evolution. Current Anthropology V 48(1), 146-153.

Torrence, R. 1983. Time budgeting and hunter-gatherer technology. In: G. Bailey (Ed.), Pleistocene Hunters and Gatherers in Europe, pp. 11-22. Cambridge University Press, New York.

Torrence, R. 1989. Tools as optimal solutions. In: R. Torrence (Ed.), Time, energy and stone tools, pp. 1-6. Cambridge University Press, Cambridge.

of London B, Biological Sciences 363.

Zimmer, C. 2008: "What is a species?" Scientific American 298 (6): 72-79.

Zimmerman, W. 1931 [1937]: "Arbeitsweise der botanischen Phylogenetik un anderer Gruppierungswissenschaften". In E. Abderhalden (Ed.), Handbuch der biologischen Arbeitsmethoden, pp. 941-1053. Berlin: Urban & Schwarzenberg.

ON THE PROBLEM OF IDENTIFYING HOMOLOGIES IN LITHIC ARTIFACTS

Gustavo Barrientos[1, 2]

(1) División Antropología. Facultad de Ciencias Naturales y Museo (FCNyM). Universidad Nacional de La Plata (UNLP). República Argentina; (2) Consejo Nacional de Investigaciones Científicas y Técnicas (CONICET); e-mail: barrient@museo.fcnym.unlp.edu.ar

Abstract

Homology is a fundamental concept in historical sciences, i.e. those disciplinary fields aimed at explaining natural phenomena in terms of causes that operated in the deep past. In archaeology, the interest for homology recognition has notoriously increased in recent years under the influence of the evolutionary paradigm. In this field of inquiry, the prevailing approach to homology is one based on the cladistic methodology. As a consequence, the homology concept lying behind most of the current phylogenetic studies on artifacts, including lithics, is a narrow one almost exclusively based on the similarity criterion. The aim of this paper is to underscore the need to enlarge the framework for homology recognition and testing in evolutionary archaeology by adopting a developmental perspective, based on the study of ontogenetic trajectories of artifacts on an individual and population basis. It is argued that the success of such an approach is highly dependent on the acquisition of a detailed knowledge about the relative contribution, on morphological variation patterns, of factors controlling the ontogeny of artifactual phenotypes, i.e. developmental homeostasis (through canalization or developmental stability) vs. phenotypic plasticity. These and closely related issues, like the assessing and quantification of the relative contribution of informational and environmental factors to the constitution of artifactual phenotypes, should integrate an agenda for future work on this topic.

KEYWORDS: HOMOLOGY, LITHIC ARTIFACTS, ONTOGENY, REPLICATIVE EXPERIMENTATION

Resumen

Homología es un concepto fundamental en las ciencias históricas, es decir, en aquellos campos disciplinarios orientados a explicar los fenómenos naturales en términos de causas que operaron en el pasado. En arqueología, el interés por el reconocimiento de las homologías se ha incrementado notoriamente en los últimos años bajo la influencia del paradigma evolutivo. En este campo de investigación, el enfoque predominante para el tratamiento de las homologías está basado en la aplicación de la metodología cladística. Como consecuencia de ello, el concepto de homología subyacente en la mayoría de los estudios filogenéticos actuales es un concepto estrecho, casi exclusivamente basado sobre el criterio de similitud. El objetivo de este trabajo es poner de relieve la necesidad de ampliar el marco para la formulación y puesta a prueba de hipótesis de homología dentro del campo de la arqueología evolutiva mediante la adopción de una perspectiva basada en el estudio de las trayectorias ontogenéticas de los artefactos, tanto sobre una base individual como poblacional. Se concluye que el éxito de este enfoque depende, en gran medida, de la adquisición de un conocimiento detallado acerca de la influencia relativa, sobre los patrones de variación morfológica, de factores que controlan la ontogenia de los fenotipos artefactuales, i.e. la homeostasis del desarrollo (a través de la canalización o de la estabilidad del desarrollo) *vs.* la plasticidad fenotípica. Estos y otros temas estrechamente relacionados entre sí, tales como la evaluación y la cuantificación de la contribución relativa de la información culturalmente heredada y cognitivamente procesada y del ambiente sobre la constitución de los fenotipos artefactuales, deben integrar una agenda para el trabajo futuro sobre este tema.

PALABRAS CLAVE: HOMOLOGÍA, ARTEFACTOS LÍTICOS, ONTOGENIA, EXPERIMENTACIÓN REPLICATIVA

Introduction

Homology is a central notion in historical sciences (Bigandt and Griffiths 2007), i.e. those disciplinary fields aimed at explaining natural phenomena in terms of long past causes (Cleland 2002). The anthropological and archaeological concern for reliability in homology identification, particularly in the context of distinguishing between homologous and analogous cultural similarities, has a very long history indeed (e.g., Galton 1889; Kroeber 1931; Binford 1968; Clarke 1968; O'Brien and Lyman 2000; Hodder 2001; see O'Brien 2010: 313-317). In fact, the clear distinction between homology and nonhomology (i.e. analogy, homoplasy) is a very important archaeological matter—despite of the archaeologist's paradigmatic orientation or bias, as Hodder (2001: 85) nicely reminds us—whenever we are interested in establishing historical connections between facts concerning past human behaviour. Moreover, in any attempt to investigate evolutionary relationships—both phylogenetic and ontogenetic—between and among artifactual phenotypes of any class (e.g., Foley 1987; Collard and Shennan 2000; O'Brien *et al.* 2001, 2002; Tehrani and Collard 2002; Jordan and Shennan 2003; O'Brien and Lyman 2003; Shennan and Collard 2005; Darwent and O'Brien 2005; Eerkens *et al.* 2005; Jordan and Mace 2005; Buchanan and Collard 2007, 2008; Riede 2009; Lycett 2010), such a distinction is not only desirable but crucial. In view of its importance, it is surprising that the identification of homologies in artifacts currently constitutes a rather underdeveloped area of archaeological inquiry (cf. O'Brien 2010: 327).

The idea advanced in this paper is that the homology concept lying behind most of the current phylogenetic studies on artifacts, including lithics, is a narrow one almost exclusively based on a similarity criterion (e.g., Lycett 2007: 544). This can be considered as the logical outcome of the cladistic methodology adopted by many of such phylogenetic studies (e.g., O'Brien *et al.* 2001, 2002; O'Brien and Lyman 2003; Buchanan and Collard 2007, 2008; Riede 2009; Scheinsohn 2009; Scheinsohn *et al.* 2009; Lycett 2010), which stresses the equivalence between homologies and synapomorphies (Patterson 1982; de Pinna 1991), and the adoption of a posterior homology assessment approach (Kleisner 2007) that uses inferred phylogenetic relationships to test homology hypotheses based on inferred monophyletic groups after the application of parsimony and congruence criteria (Patterson 1982, 1988; de Pinna 1991; Haszprunar 1992; Brower and Schawaroch 1996; Sluys 1996). Under this approach, homology becomes the end-product of a phylogenetic analysis rather than something that needs to be thoroughly investigated—using a variety of criteria like comparative morphological studies, topology, connectivity, ontogeny, and functional anatomy (Kearney and Rieppel 2006)—before (or besides) cladogram construction and evaluation. An obvious consequence of this practice is that many potentially important and useful homologous characters not similar in position, structure, or function tend to remain virtually undetectable.

My aim here is not so much to criticize the assumptions and methodology of current phylogenetic studies on material culture as to underscore the need to enlarge the framework for homology recognition and testing in evolutionary archaeology. This goal is in line with recent claims in systematics and philosophy of biology which advocate for at least some causal grounding for homology hypotheses beyond mere congruence (Rieppel and Kearney 2002; Kearney and Rieppel 2006; Love 2007; Winther 2009; Assis and Rieppel 2011), as well as with the main tenets of the emerging research program that is looking for a unified approach to homology through the coherent integration of ontogeny and phylogeny (Laubichler 2000; Rieppel 2005; Wagner 2007; Wagner *et al.* 2007; Ereshefsky 2009). In order to achieve the proposed aim I will: (1) briefly review the current literature about the meaning of homology in evolutionary biology and the different available criteria to recognize homologous characters, (2) present and discuss the assessment criteria for homologies used by most of the phylogenetic studies on artifactual variation carried out over the last 10 years, (3) propose complementary ways of searching for homologies in artifacts, highlighting the advantages and problems associated with their implementation, and (4) sketch the lines of research that would be beneficially explored in order to gain knowledge about patterns of artifactual variation and about the mechanisms involved in the evolution—both ontogenetic and phylogenetic—of the artifactual phenotypes.

The meaning of homology in evolutionary biology and the different ways of identifying homologous characters

To try to summarize all the points of contention in the current debate surrounding the homology concept(s) is a very difficult task, due to the inherent complexity of the subject matter and to the considerable bulk of literature on the issue that has been produced in recent years (for comprehensive and critical reviews see, among others, Patterson 1982; Rieppel 1988; Donoghue 1992; Hall 1994, 2003; Lauder 1994; Bock and Cardew 1999; Brower 2000; Brigandt 2003; Cracraft 2005; Kearney and Rieppel 2006; Griffiths 2007; Kleisner 2007; Wake 2003; Ereshefsky 2009). In this section I will only concentrate on those aspects relevant to my discussion, namely the restriction of the meaning of homology operated in cladistics—with all its conceptual and practical consequences—, and the theoretical and methodological alternatives existing today in evolutionary biology to tackle the problem of homology recognition.

Among the many conceptualizations of homology proposed during the last five decades, the two most widely accepted nowadays are the taxic and the developmental approaches (Griffiths 2006; Kleisner 2007; Ereshefsky 2009). In the taxic conception, which was embraced by the so-called pattern or transformed cladistics (*sensu* Beatty 1982) or, more simply, Cladistics (with a capital C; Williams *et al.* 2008: 85) over the last 30 years, characters in different taxa are homologous if those taxa possess the character by descent from a common ancestor (Griffiths 2007), thus implying a hierarchy of groups and the formulation of hypotheses of monophyly (Patterson 1982, 1988; Stevens 1984; de Pinna 1991). For the developmental approach, two characters are homologous if they share the same set of developmental constraints (Wagner 1989a), which conforms a developmental module causing the stable production of a homologue (Wagner 1996). Such developmental module would consist of a gene regulatory network composed by the genes and their corresponding interactions (Abouheif 1999; Bolker 2000; Wagner 2007; Wagner *et al.* 2007; Kuratani 2009).

The taxic approach to homology has its formal origins in the work of Patterson (1982), who draws on the distinction made by Eldredge (1979) between taxic and transformational approaches to process analysis in evolution. Patterson (1982), following the propositions of several authors (e.g., Wiley 1975, 1976; Bonde 1977; Platnick and Cameron 1977; Szalay 1977; Cracraft 1978; Nelson 1978; Gaffney 1979; Platnick 1979; Eldredge and Cracraft 1980; Nelson and Platnick 1981; see discussion in de Pinna 1991), explicitly established an equivalence in meaning between homology and synapomorphy (i.e. a derived or specialized character shared by two or more taxa that originated in their last common ancestor), which is a fundamental concept in the logical framework of cladistics. The proposed equivalence was explicitly acknowledged by Nelson and Platnick (1981: 137) and de Pinna (1991: 369) as a restriction in the meaning of the term 'homology', whose philosophical justification resides in the fact that such restriction implies a greater logical specificity and a greater empirical content of the concept, and hence an increased scientific value. From a methodological standpoint, de Pinna (1991: 371) states that "by equating all derived similarities with synapomorphy, it is possible to detect a common pattern of internested attributes". It is precisely under such framework, which operationally makes all characters equivalent to hypothesized synapomorphies, that the procedure of finding out a common pattern among taxa can be performed (de Pinna 1991).

In the taxic approach that pervades current cladistics, a homology—considered as a mental construct or abstraction (Patterson 1982, 1988, Nelson 1989; see discussion in Kleisner 2007: 328)—has always the status of a hypothesis. In consequence, the main effort is put in the process of hypothesis testing, which can be accomplished with the aid of the three tests proposed by Patterson (1982): similarity, conjunction, and congruence. It is intended that the results of these tests can discriminate different categories of homology and homoplasy (nonhomology) (Patterson 1988: 604). For Patterson (1982, 1988), similarity—concerning topographic, histological, developmental, and compositional attributes—is deemed the traditional criterion of comparative morphology. Since nonhomology also implies similarity, it is considered an insufficient and weak tool to tests a hypothesis of homology, but useful enough to validate a hypothesized homology as worthy of further examination because, as Cracraft (1981: 25) rightly pointed out, "similarity is the factor that compels us to postulate homology" (see also Stevens 1984: 403). Conjunction, devised by Patterson (1982) as a test capable of disproving homologies as anatomical singulars, excludes the presence of two homologues within a single organism, i.e. if two putative homologues are found together in one organism, they logically cannot be homologous (Rose and Lauder 1996). Congruence, finally, is considered the most decisive test of homology (Patterson 1988: 605). It depends on the equivalence between homology and synapomorphy, and is based on the principle of parsimony (Patterson 1982: 38). In testing a hypothesis of homology by congruence, the distribution of the target feature is confronted against the distribution of other putative homologies: a proper homology will define a group that is congruent with those specified by other homologies. Comparatively, then, congruence test is the most powerful of the three assessment procedures proposed by Patterson (1982), and purportedly the only one capable of accurately discriminate between homology and homoplasy.

In summary, it can be said that the taxic definition of homology attaches great importance to the systematic requirement for monophyletic trees, adjudicating a minor or a secondary significance to anatomical and developmental facts like position or embryological origin (Amundson 2001). Under the taxic approach adopted by cladistics, the determination of homologies takes the form of a at least two-step procedure (cf. Brower and Schawaroch 1996): the first step consists in the identification of a putative, i.e. hypothetical, homology between characters or character states present in two or more taxa (primary homology, *sensu* de Pinna 1991), which is inferred on the basis of some a priori criteria, mainly similarity; the second step consists in the testing of the homology hypothesis by means of a phylogenetic analysis based on the principles of congruence and parsimony. The homology propositions that successfully pass such tests are then called secondary homologies or synapomorphies (de Pinna 1991).

In opposition to the taxic approach, the developmental stance to homology centres its attention on ontogeny rather than on phylogeny (Ereshefsky 2009). As presented by Wagner (1989a, 1989b, 1999), the so-called 'biological homology concept' emerged within this framework as an approach to homology interested in generative biological mechanisms (i.e. processes) rather than in purely genealogical relations.

Starting from the premise that homology is a relation between parts of the body and drawing on ideas contained in the work of Roth (1984) and of early and middle 20th Century biologists like H. Spemann, F. Baltzer and H. Kroeger, Wagner (1989a: 62) provided a preliminary definition of biological homology which states that

> "Structures from two individuals or from the same individual are homologous if they share a set of developmental constraints, caused by locally acting self-regulatory mechanisms of organ differentiation. These structures are thus developmentally individualized parts of the phenotype."

It is followed from this definition that the homology relation cannot be meaningfully predicated of all conceivable features, but only of those developmentally constrained morphological patterns (Wagner 1989a: 66).

Wagner (1989b) considers that the research program on the biological basis of homology should approach the evolution of morphological characters at two different levels. The first one refers to character modification within a definite framework of developmental constraints, whereas the second one refers to the evolutionary modifications of such constraints. The need to deal with the problem of homology at two different levels has also been acknowledged by Laubichler (2000), who argues that homology should be explained with reference to both, development and (phylogenetic) evolution, since development is the proximate cause of morphological characters and phylogenetic evolution implies either organic transformation or stability.

Functional to the growth of the developmental approach to homology—and to the entire field of evolutionary development or 'evo-devo'—has been the view, adopted by many during the last 15 years, that morphological characters at the organismic level may be considered the phenotypic units of evolutionary transformation, units that can be meaningfully compared more or less independently one from another in an interspecific context (Wagner 1996). Such units are called modules (Wagner 1996; Raff 1996; Carroll *et al.* 2001; Schlosser and Wagner 2004; Wagnet *et al.* 2007; Klingenberg, 2008; Kuratani 2009), and it is intended that they conceptually fill the gap between genes and phenotypes by providing a possible link between both domains (Kuratani 2009: 61). In this connection, different authors have proposed that the developmental modules, considered as units of evolvability (Laubichler 2000; Brigandt 2007; cf. Ereshevsky 2009), consist of a gene regulatory network composed by the genes and their respective interactions (Abouheif 1999; Bolker 2000; Wagner 2007; Wagner *et al.* 2007; Kuratani 2009).

Within the framework of the developmental approach to homology, Wagner (1999: 125) has stipulated a research program for testing the biological homology concept based on a combination of comparative, genealogical, and experimental methods. Such research program entails the following steps:

> "(1) identifying of two or more putative homologues in a clade; (2) determining the phylogenetic distribution of the putative homologues; (3) describing the intra- and interspecific variation patterns of each putative homologue; (4) describing the development of each putative homologue, and determining if modes of development and distribution of homologues are phylogenetically congruent; and (5) providing and testing a model of how differences in modes of development between putative homologues effect differences in variational tendencies."

As it has been widely debated (e.g., Roth 1988; Brigandt 2002, 2003, 2007, 2008; Hall 2003, 2007; Wake 2003; Grant y Kluge 2004; Cracraft 2005; Ghiselin 2005; Kearney and Rieppel 2006; Griffiths 2006, 2007; Bigandt and Griffiths 2007; Kearney 2007; Kleisner 2007; Rieppel and Kearney 2007; Matthen 2007; Winther 2009; Assis and Rieppel 2011), the taxic and developmental approaches to homology definition and recognition are source of both enlightenment and shortcomings depending on the application context. In part to overcome this situation, some scholars are now advocating for a unified approach to homology, one that coherently integrates ontogeny and phylogeny into a single operative framework (Laubichler 2000; Rieppel 2005; Wagner 2007; Ereshefsky 2009). Recently, Ereshevsky (2009) has given an account of the rationale for such integration that is based on three affirmations: (1) *phylogeny is essential for establishing the identity conditions of a homologous character*; (2) *ontogeny is essential for explaining variation and similarity among the character states of a homologue*; and (3) *developmental factors need to be transmitted across generations*. The success of this ongoing undertaking remains to be judged [for a pessimistic view about this whole enterprise, see Maynard-Smith (quoted in Tautz 1998: 17); Wake (2003); Griffiths (2007)], but the logic behind the effort is certainly sound.

From the precedent review it becomes clear that for to get a deeper understanding of evolution occurring at different levels along a hierarchy of entities and processes, there is an urgent need of integrating pattern and process in a coherent way. In this endeavour, the role of homology as a pivotal concept cannot be overemphasized. This alone justifies the importance of reaching a trade-off between specificity and generality in homology definition, and of developing an integrated and coherent methodological approach for homology recognition and testing. This is something that is valid for biology and for archaeology as well.

The treatment of homology in evolutionary archaeology: A critical appraisal

As stated above, the prevailing approach in evolutionary archaeology to the homology identification problem is one based on the cladistic methodology. The rationale behind this fact resides in the current emphasis on phylogenetic reconstruction (i.e. what causes what and in what order; O'Brien and Lyman 2003) as a way to tackle the issue of cultural transmission and artifactual evolution (i.e. if things evolve in a nonrandom way, they can be properly studied by the use of cladistics; O'Brien and Lyman 2003). In this context and as long as cladistics became the most commonly used phylogenetic methodology in biological systematics over recent decades, most evolutionary-oriented archaeologists adopted it in order to investigate historical connections concerning artifactual data (e.g., O'Brien *et al.* 2001; Lyman and O'Brien 2005; contributions in O'Brien 2008; Lycett 2010). As a consequence, cladistics virtually replaced seriation, which was the preferred approach in evolutionary archaeology in its early days (Dunnell 1970; Teltser 1995; Lyman *et al.* 1998).

As depicted by Buchanan and Collard (2007: 368) and Lycett (2007: 544; 2010: 210-211), archaeological applications of cladistics proceed along the same series of steps used in biological systematics (McLennan and Brooks 2001). Under this approach, homologies are considered as hypotheses about historical relationship between characters or character states present in two or more analytical units referred to as Operative Taxonomic Units (OTUs). Such hypotheses are mainly formulated on the basis of a similarity criterion and are tested by means of the inferred phylogenetic relationships after analysis completion (Lycett 2007: 544) (i.e. posterior homology assessment approach; *sensu* Kleisner 2007).

It is important to note that archaeological cladistics uses both qualitative and quantitative (morphometric) data, either separately or in combination (e.g., O'Brien *et al.* 2001; Buchanan and Collard 2007, 2008; Lycett 2009a, 2009b, 2010). In morphometrics, particularly in landmark or geometric morphometrics, 'homology' primarily refers to "correspondence of point(s) (or measurement) across the range of lithic forms in a given analysis" (Lycett 2009a: 88), not necessarily implying "common evolutionary (i.e. genealogical) ancestry and/or common developmental pathways" (Lycett 2009a: 88; see also Buchanan *et al.* 2007: 284-285, and Lycett and Chauhan 2010: 16-18). It is worth reminding that the primary concern of morphometrics is to ensure comparability between cases in form (both, shape and size) analysis in order to proportionate meaningful information about variation within a collection of specimens, and not strictly to reveal or assess genealogical relationships (O'Higgins 2000: 105). In fact, the degree to which either 2D or 3D landmark configurations or even distances between points sampled from a biological form relate to truly homologous structures is, usually, difficult to estimate owing to the allowed combination, in geometric morphometrics, of different kinds of reference points (i.e. Type I, Type II, and Type III landmarks; Bookstein 1991, 1997), each of them carrying a distinct 'homology signal' (see discussion in Cardillo 2010: 327-328). Nevertheless, if each point is clearly defined and consistently identified, the comparability of landmark configurations or measurements between specimens in a sample is guaranteed. In any case, it is clear that in morphometrics—like in cladistics—homology claims always adopt the form of a priori putative statements (O'Higgins 2000: 106; Strauss 2010: 73), which, in this case, remain basically untested throughout the entire analysis.

Two main criticisms can be raised against the widespread treatment of the homology problem in evolutionary archaeology, at least as it was succinctly reviewed here. The first one refers to the use of the similarity criterion as the only source of hypotheses about homology. The second one refers to the current lack of emphasis on the development of a sound methodology for detailed character analysis in order to independently test homology assertions, as well as to discover unexpected homologous relationships between morphological traits.

Despite the fact that morphological similarity—i.e. the extent to which one form resembles another form, as determined according to some specified set of criteria and procedures—is often considered the prerequisite of a homology hypothesis (Cracraft 1981: 25; Stevens 1984: 403), and even "the very phenomenon that the systematic method is supposed to account for" (de Pinna 1991: 377), it is clear that traits similar in terms of size, shape or function may be non-homologues and traits dissimilar in terms of such properties may be homologues (Rieppel and Kearney 2002: 63). Indeed, morphological similarity as a criterion for homology identification may be potentially misleading due to the fact that evolutionary modifications can cause a loss of morphological resemblance of homologous structures, while reasons other than common ancestry (e.g., convergent evolution) can make two non-homologous structures appear quite similar thus complicating the recognition of truly homologous structures (Lao and Tomoyasu 2011: 232).

It is relevant to introduce here the key distinction between the terms 'similarity' and 'sameness'. To put it simply, homology is a relation of sameness not similarity (Müller 2003; Ghiselin 2005; Ereshefsky 2009). Sameness, the quality or condition of being the same or identical (Noonan 2009), was first brought in by Owen (1843) as the criterion to distinguish between homology, i.e. the presence of the same parts in every variety of form and function, and analogy, i.e. similarity of parts in their functional adaptations. As it is understood, structures in different taxa are proposed to be homologous only if it is thought that the *same* structures—in some specified sense—were present in their common ancestor, irrespective of their degree of similarity or resemblance (Puelles and Medina 2002: 243). From a logical point of view, we can have sameness with similarity (homology), sameness without similarity (homology), and similarity without sameness (analogy, homoplasy). It follows that similarity is neither a necessary nor a sufficient criterion for homology identification. The cladistic approach is undoubtedly appropriated to discriminate between homology and homoplasy whenever similarity coincides with sameness, but it has little to offer in cases in which sameness and similarity do not co-occur (by the virtual absence of the latter).

An important limitation of the cladistic methodology, as far as it concerns to the homology problem, resides in its disregard in performing detailed, in-depth assessment of character hypotheses—namely primary conjectures of homology—beyond the test of congruence, which specifically does not deal with character analysis in itself (Rieppel and Kearney 2002: 59). The generalized adoption of such methodology by archaeological research oriented to investigate variation patterns in artifactual phenotypes, in some way precludes the implementation of a more sophisticated approach to the treatment of homology than that that is usually found in the current literature on the subject (Lycett 2010; O'Brien 2010). Just like in the case of biological systematics, what evolutionary approaches in archaeology need to develop is a sound methodology aimed to evaluate characters in terms of their developmental and/or functional causes (Rieppel and Kearney 2007; Rieppel 2008). Some ideas about the way to overcome present limitations in relation with this problem shall be discussed in the next sections of this paper.

Towards an intergrative approach in the study of homologies in lithic artefacts: The rationale, the strategy, and the problems

As recent theoretical developments in biological systematics suggest, a unified approach to homology—i.e. one integrating ontogeny and phylogeny into a single interpretive and operative framework (Laubichler 2000; Rieppel 2005; Wagner 2007; Wagner *et al.* 2007; Ereshefsky 2009)—represents a seemingly viable project aimed to address problems that neither the taxic nor the developmental approaches alone could satisfactorily solve (e.g., how to accommodate the multiple, and often competing perspectives and research agendas on homology?; how to provide solid causal grounding to our primary homology conjectures or character statements?; how to understand the phenomenon of hierarchical disconnect, i.e. the fact that non-homologous factors at one level can cause homologous traits at a higher hierarchical level and, conversely, homologous factors can cause nonhomologous characters?; Rieppel and Kearney 2007; Rieppel 2008; Ereshevsky 2009).

The main tenet of this paper is that in order to enlarge the framework for homology recognition and testing in evolutionary archaeology, it is crucial to engage in a synthesis similar to that is now ongoing in the field of biological systematics. It is significant to note that, although not directly addressing the problem of homology identification and validation, some scholars (Riede 2006, 2008; Shott 2008, 2010; Lycett 2010) have already started to call attention on the importance of integrating ontogenetic aspects of lithic and other artifacts (e.g., reduction sequences) with patterns of phylogenetic evolution, in order to get a better understanding of the evolutionary process underwent by such material entities. At present, however, the precise way in which the advocated synthesis may proceed is less than clear.

A fundamental step towards a unified approach to homology in the evolutionary study of artifacts should be, I would argue, to establish equilibrium between phylogenetic and character analyses since, as it was already mentioned, there is a marked trend to lay emphasis on the former at the expense of the latter. Homology hypotheses should be testable in other ways than just on the basis of inferred monophyletic groups after the application of the congruence criterion (Rieppel and Kearney 2002: 63). The test of morphological characters can indeed be carried out using the main criteria of homology proposed in the early 1950's by the German systematist Adolf Remane: a) topological equivalence (criterion of 'sameness of position'), b) special quality of structures, and c) linkage by intermediate forms (Wenzel 1992: 369; Rieppel and Kearney 2002: 59). Topological equivalence, the first and foremost criterion of homology, is based on the idea that homologous structures share similar locations and connections with adjacent parts; special quality designates any single characteristic or trait contributing to the distinctiveness of a structure, in such a way that homologues tend to present the trait more likely than non-homologues; the existence of intermediate forms, finally, imply that homologous structures tend to be more similar early in ontogeny than later, owing to the fact that intensive evolutionary modification can cause a loss of morphological similarity of homologous structures in adult (i.e. fully functional) forms (Moczek 2008: 436-437). At the core of Remane's proposal lies a probabilistic approach to homology (Laubichler 2000: 780): a homology hypothesis increases in likelihood when more than one criterion is fulfilled. All these criteria are applicable to data coming from such disparate sources as comparative anatomy, comparative fossil anatomy, comparative embryology, comparative physiology and biochemistry, and comparative ethology (Wenzel 1992; Rieppel and Kearney 2002; Brigandt 2003), and deal—particularly the first and third—with the causal processes of ontogeny. And ontogeny matters because morphological homologues are complex structures that change through time (Griffiths 1999).

The endeavour of testing homology hypotheses concerning lithic artifacts would benefit from the systematic and comparative study of large collections of artifacts from archaeological contexts but, above all, from the comparative study of collections of experimentally produced tools and debitage. By means of replicative experiments (despite their potential pitfalls; see critical appraisals in Morrison 1994 and Koerner 2011), the ontogeny of lithic artifacts can be approached in an analytic way similar to that of developmental biology. An alternative to experimentation would be the use of ethnographic and ethnoarchaeological data about the life-histories of different artifact classes. However, it must be realized that there is a general paucity of ethnographic information about lithic tool production and use (Shott and Sillitoe 2005: 654). Moreover, while this kind of technology was still in use—at least occasionally—until recently (i.e. 1980's) in different regions around the world (e.g., Australia, Central America, Amazonia, Eastern Africa, and the New Guinea Highlands), it is virtually impossible nowadays to find people engaged in chipped stone production on a regular basis due to the massive adoption of metal tools and weaponry almost everywhere (Sillitoe and Hardy 2003: 555). This situation clearly hampers the implementation of an ethnoarchaeological approach to the subject.

Replicative experiments of specific lithic reduction sequences (Flenniken 1984; Whittaker 1994 and literature cited therein) possess the advantage of providing a good context, at a relative low cost, for the testing of homology hypotheses following Remane's criteria in highly controlled conditions, allowing for the reiterative deployment of procedures and the statistical analysis of resulting data. It cannot be overemphasized, however, that the success of the proposed strategy depends on the resolution of a series of questions related to the process and pattern of artifactual phenotypic variation that are begging for a more clearer understanding than that is available today.

I would argue that a developmental approach to homology recognition and testing may represent a practicable undertaking only if it can be demonstrated that: a) the life-history of different artifact classes is capable of being assimilated—to a certain extent—to the biological concept of ontogeny in a relevant or not merely impressionistic way; and b) the artifactual phenotypes exhibit a certain degree of constancy (Waddington 1957) at the individual and population levels. In other words, it seems clear that only if lithic artifacts truly develop, and in a non-random or idiosyncratic fashion, an ontogenetic or developmental approach to homology is feasible. In the remainder of this section I shall briefly discuss these pressing issues.

The first problem that cries for clarification is the extent to which the concept of ontogeny is relevant to describe the life-history of lithic artifacts. Ontogeny can be broadly defined as the history of structural change affecting a unity (whatever it may be), without the loss of the organization which allows that unity come to existence (Maturana and Varela 1987: 74). Ontogeny form part of the more general phenomenon of intarvariation, which consists in the internal variation of a system as a consequence of the instability of structural units over time (cf. Schreider 1960). Organismal units express such intravariation through the fundamental processes of growth, development, and senescence, all these being fundamental causal forces contributing to phenotype construction along life cycle. Phenotype construction, principally during its early phases, is inherently morphogenetic, with each event dependent on previous ones and contextualized by collateral ones (Salthe and von Sternberg 2009).

Lithic artifact production and use are reductive phenomena (Andrefsky 2005: 30, 34-38; Clarkson 2007: 32). This implies that the general process affecting most artifact classes (either tools or debitage) along their life-history trajectories (i.e. the succession of events extending "from the first flake removed from a core though any potential resharpening and usewear that occurs prior to final discard"; Lycett 2010: 227; see also Andrefsky 2009: 67) is decrease instead of growth (e.g., a biface cannot be bigger than the blank from where it was produced, becoming even smaller as the reduction sequence advances). This general statement, however, may not be true in all cases: e.g., unused flakes resulting from the reduction of a core or other artifacts do not undergo further decrease until discard, which can be immediate. The general norm of decrease or stability (either by sharpening, wear, or discard) in the life-history of lithic artifacts introduces a significant difference with the ontogeny of organismal units, which tend to grow by increase in the size and/or the number of components (Cameron 2002). A further discrepancy comes from the fact that organismal units undergo a process of increasing differentiation as growth progresses. Such differentiation involves diversification in the form, function and hierarchical organization of components. Clearly, lithic artifacts do not engage into a process of internal differentiation, at least in the same sense that cells, tissues or whole organisms do. Chipped stones are more like some hard components of the phenotype of certain organism (e.g., mollusk valves) in that they depend on the growth, differentiation, and maturity of other components of the phenotype in order to come into existence and achieve a definite form. Stone artifacts experience a sort of mechanical (as opposed to organic) differentiation, in the sense of becoming increasingly segmented (or at least susceptible of being segmented) into more or less identifiable parts (e.g., an endscraper as it is shaped into a piece with a sharp working edge, a proximal hafted end, and two lateral edges).

Another fundamental process contributing to morphogenesis is development. In general terms, development is progressive change, a process of fine-tuning enhancement through the differential promotion of latent informational constraints (i.e., sites or regions that might assume more than one configuration), followed by the ensuing fixation of some of such constraints (Salthe 2010: 358-359). The 'canonical developmental trajectory' of different systems (Salthe 1993) typically runs from immaturity (dominated by morphogenesis) through maturity (in which form is definitive for the kind of system) to senescence (in which form increasingly accumulates distorting marks as a result of interactions with the environment) (Salthe 2010: 60). Biological development—a phenomenon intimately associated with growth—is mainly conceived as the attainment of an increasing functional ability (maturity) along the ontogenetic process (Cameron 2002: 10), functional ability being measured in relation with some particular function or set of functions. Chipped stones (or at least some of them) 'develop' in the sense that they experience progressive change (i.e. decreasement and shape modification). The question that arise, however, is if lithic tools really reach something like a mature (i.e. fully functional) or definitive form. On the one hand, from the perspective of the so-called 'teleological model' (Bleed 2001)—i.e.

that embraced by the *châine opèratoire* school as it was characterized by Dibble (1995), Bleed (2001), and Shott (2003)—reduction sequences are considered as more or less prolonged series of actions aimed to a somewhat predetermined goal or end-product, following a sort of mental template. From this standpoint, tools would really attain a mature stage in which their functional ability or performance is the highest. On the other hand, the rival 'evolutionary model' (Bleed 2001) emphasizes the rather contingent and situational nature of lithic technology, understanding stone production as a continuous and "expanding array of alternatives defined by intervening options and outcomes" (Bleed 1991:20) which makes unlikely, in most cases, the existence of predictable end-results. Under this perspective, the very idea of a mental template is put into question (e.g., Bleed 2001; Davidson 2002; cf. Costa 2010), as it is the proposition of a clearly identifiable 'mature' stage in the life-history of most tools. This contentious issue is far from being resolved, but it seems prudent to assume that at least certain tools—likely those in which function and selection for performance impose very narrow constraints on form and morphogenetic process—indeed attain an state that could be properly considered as 'mature' immediately before the point of first usage, while other less functionally and morphologically less constrained tools change in a apparently undirectional way. I shall return later on this point when discussing the concept of 'canalization'.

Senescence or aging, finally, designates irreversible changes—assessed by means of specific biomarkers (Fox 2007)—in structure and function (generally involving impairment or loss of a function) of a whole organism or part of an organism that occur after maturity as a result of time-dependent processes (Medina 1996). It is an observational fact that lithic edges eventually become worn and blunt, then needing some amount of restoration for further use with the same or a closely related function (Andrefsky 2009). This is understood by some as representing senility or aging following first usage (Lycett 2010: 227 footnote nº 2; Shott 2010: 275). The indicators of such process (analogous to biomarkers) are usewear and resharpening. It must be noted that the strict application of the concept of senescence to stone tools is logically reliant on the acceptance—at least partially—of the tenets of the teleological model about lithic reduction. Tool must mature (i.e. to reach a definitive and fully functional form) in order to age. If tools do not mature, then they cannot age but simply change. Instances of tools that seems to fit well into the developmental model are most of the North American bifaces (e.g., projectile points), which go through a clearly differentiated production phase and through a subsequent use-life phase (i.e. the service life of one tool; Shott and Sillitoe 2005: 654) (Wilson and Andrefsky 2008: 87; Andrefsky 2009: 69). Examples of the entangled production-use model which implies a continuously changing form resulting from successive phases of use and resharpening are flake knives (Andrefsky 2009: 69), Australian endscrapers (Clarkson 2005), and most bifaces from areas of the world different than North America (Wilson and Andrefsky 2008: 87; cf. Clarkson 2002).

The second problem refers to the degree of constancy exhibited by those artifactual phenotypes that indeed develop. The concept of phenotypic constancy is linked to the notion of developmental homeostasis, i.e. "the ability of an organism to maintain a more buffered series of developmental pathways resulting in increased phenotypic uniformity of individuals in a population" (Etges 1989: 189). Developmental homeostasis ensures phenotypic constancy in the face of genetic, environmental, and developmental variation (Waddington 1957), and is controlled by two main processes: canalization and developmental stability (Debat and David 2001). Canalization designates the process by which phenotypic variation is reduced, at the population level, by developmental mechanisms (Stearns *et al.* 1995). Canalization buffers development to reduce the amount of variation potentially introduced by genetic and environmental factors, thereby ensuring phenotypic constancy in populations (Clarke 1998; Debat *et al.* 2000). Developmental stability, in turn, is the ability of an organism to maintain a stable development under a particular set of conditions due to the existence of mechanisms that reduce variability caused by accidents or random perturbations, i.e. noise (Clarke 1998; Van Dongen and Lens 2000). Canalization reduces phenotypic variation among individuals whereas developmental stability reduces variation within individuals (Etges 1989). The lack or low degree of homeostasis is called phenotypic plasticity (Etges 1989), which is defined either as the ability of a genotype to express different phenotypes across a range of environments (Liefting *et al.* 2009), or the ability of an organism to alter its physiology, morphology, and development in response to changes in the environment (Callahan *et al.* 1997). Phenotypic plasticity usually varies between habitats as a function of the degree of spatio-temporal heterogeneity (Liefting *et al.* 2009).

We can expect that the possibility of approaching homologies in lithic artifacts by means of the study of developmental patterns at the individual and population levels, heavily depends on the degree of developmental homeostasis in the ontogeny of artifactual phenotypes: i.e. the lesser the degree of developmental homeostasis, and hence of phenotypic constancy, the lesser the chances of obtaining useful information about potentially homologous traits. Nevertheless, a proper assessment of the relative contribution of developmental homeostasis and phenotypic plasticity to the patterns of artifactual variation requires a deep understanding—and some degree of quantification—of the influences of informational (i.e. genetic, epigenetic, behavioral, symbolic; Jablonka and Lamb 2005) and environmental factors (including the interaction between both kinds of causes) on the expression of the artifactual phenotypes. At the present time we notoriously lack such understanding and analytical sophistication, so an agenda for further work on this subject should unavoidably include these critical topics.

Concluding remarks

To enlarging the framework for homology recognition and testing at the level of artifactual phenotypes is a much relevant goal in evolutionary archaeology, as long as this field aspires to obtain a deeper understanding of evolution occurring at different levels along a hierarchy of entities and processes involving the human clade. Current emphasis on cladistics as the almost unique context in which homology hypotheses are formulated and tested should not hinder the exploration of other sources of information about homologous traits in lithic artifacts. In this vein, I advocated here for a more equilibrated approach to the homology problem, one integrating phylogenetic and ontogenetic perspectives into a coherent research program. However it must be realized that, in order to make of this project a viable effort, a great deal of knowledge about the many factors influencing the patterns of intervariation and intravariation in artifactual phenotypes ought to be achieved. Chief among the issues that need to be urgently addressed is the assessment and quantification of the differential contribution of those factors involved in the construction of artifactual phenotypes, namely the culturally inherited and cognitively processed information, the environment, and the interaction between both factors. In this sense, one major objective of evolutionary archaeology should be to identify the causes of the phenotypic variance and covariance within and between life-history traits in artifacts, as well as the relative contribution of such sources to the total phenotypic variance within each trait. This might aid to formulate explicit models about the way in which the phenotypic variance of artifactual traits can be partitioned. The second major problem to address is the evaluation of the degree of developmental homeostasis in the ontogeny of artifactual phenotypes, particularly of those artifact classes that undergo a true developmental process as it was defined above. The pertinence of an experimental program to attack the problem of homology recognition and validation in lithic artifacts seems to be contingent on the demonstration that, at least for certain artifact classes, the morphogenetic process is constrained to a certain degree, i.e. the amount of equifinality is relatively low. However, it has to be remembered that due to the phenomenon of hierarchical disconnect, even non-homologous developmental pathways can cause the same homologous morphology (Abouheif 1999; Ereshevsky 2009). All these problems deserve to be thoroughly investigated, both theoretically and empirically.

As a final thought, I want to mention that different authors (e.g., Binford 1968; Dunnell 1971, 1978) have stated that the terms we use to create our data directly establish a determination over the kind of operations or procedures that we can subsequently execute on those data. In this sense, it might be necessary to rephrase many of the linguistic expressions already in use in lithic analysis into terms or expressions usually employed in organismal evolutionary literature, after assessing for adequacy and relevance (e.g., life-history for ontogeny, standardization for canalization). This may aid to reduce ambiguity and imprecision in the communication of ideas and results but, above all, to set the problem in a proper and long-established frame of reference.

Acknowledgements

To the convenors of the meeting "*Arqueología Argentina y el Legado de Charles Darwin*" (Argentine Archaeology and the Legacy of Charles Darwin) and editors of the present book, Hernán Muscio and Marcelo Cardillo, for their kind invitation to participate in both instances. To Luciana Catella for her comments on a draft of this paper.

References

Abouheif, E. 1999. Establishing homology criteria for regulatory gene networks: Prospects and challenges. In Bock, G. R. and G. Cardew (eds.); *Homology*: 207-225. Novartis Foundation for Gerontological Research Symposium 222. Chichester, John Wiley & Sons.

Amundson, R. 2001. Homology and homoplasy: A philosophical perspective. In *Encyclopedia of Life Sciences*. Nature Publishing Group. http://www.els.net. 12/04/2010.

Andrefsky, Jr., W. 2005. *Lithics: Macroscopic Approaches to Analysis*, Second Edition. Cambridge, Cambridge University Press.

Andrefsky, Jr., W. 2009. The analysis of stone tool procurement, production, and maintenance. *Journal of Archaeological Research* 17: 65-103.

Assis, L. C. S. and O. Rieppel. 2011. Are monophyly and synapomorphy the same or different? Revisiting the role of morphology in phylogenetics. *Cladistics* 27: 94-102.

Beatty, J. 1982. Classes and cladists. *Systematic Zoology* 31: 25-34.

Binford, L. R. 1968. Archaeological perspectives. In Binford, S. R. and L. R. Binford (eds.); *New Perspectives in Archaeology*: 5-32. Chicago, Aldine.

Bleed, P. 1991. Operations research and archaeology. *American Antiquity* 56:19-35.

Bleed, P. 2001. Trees or chains, links or branches: Conceptual alternatives for consideration of stone tool production and other sequential activities. *Journal of Archaeological Method and Theory* 8: 101-127.

Bock, G. R. and G. Cardew (eds.). 1999. *Homology*. Novartis Foundation for Gerontological Research Symposium 222. Chichester, John Wiley & Sons.

Bolker, J. A. 2000. Modularity in development and why it matters to Evo-Devo. *American Zoologist* 40: 770-776.

Bonde, N. 1977. Cladistic classification as applied to vertebrates. In Hecht, M. K., P. C. Goody and B. M. Hecht (eds.); *Major Patterns in Vertebrate Evolution*: 741-804. New York, Plenum Press.

Bookstein, F. L. 1991. *Morphometric Tools for Landmark Data: Geometry and Biology.* Cambridge, Cambridge University Press.

Bookstein, F. L. 1997. Landmark methods for forms without landmarks: Morphometrics of group differences in outline shape. *Medical Image Analysis* 1: 225-243.

Brigandt, I. 2002. Homology and the origin of correspondence. *Biology and Philosophy* 17: 389-407.

Brigandt, I. 2003. Homology in comparative, molecular, and evolutionary developmental biology: The radiation of a concept. *Journal of Experimental Zoology (Molecular and Developmental Evolution)* 299B: 9-17.

Brigandt, I. 2007. Typology now: Homology and developmental constraints explain evolvability. *Biology and Philosophy* 22: 709-725.

Brigandt, I. and P. E. Griffiths. 2007. The importance of homology for biology and philosophy. *Biology and Philosophy* 22: 633-641.

Brower, A. V. Z. 2000. Homology and the inference of systematic relationships: Some historical and philosohical perspectives. In Scotland, R. and R. T. Pennington (eds.); *Homology and Systematics: Coding Characters for Phylogenetic Analysis*: 10-21. London, Taylor & Francis.

Brower, A. V. Z. and V. Schawaroch. 1996. Three steps of homology assessment. *Cladistics* 12: 265-272.

Buchanan, B. and M. Collard. 2007. Investigating the peopling of North America through cladistic analyses of Early Paleoindian projectile points. Journal of Anthropolological Archaeology 26: 366-393.

Buchanan, B. and M. Collard. 2008. Phenetics, cladistics, and the search for the Alaskan ancestors of the Paleoindians: A reassessment of relationships among the Clovis, Nenana, and Denali archaeological complexes. *Journal of Archaeological Science* 36: 1683-1694.

Buchanan, B., E. Johnson, R. E. Strauss, and P. J. Lewis. 2007. A morphometric approach to assessing Late Paleoindian projectile point variability on the Southern High Plains. *Plains Anthropologist* 203:279-299.

Callahan, H. S., M. Pigliucci, and C. D. Schlichting. 1997. Developmental phenotypic plasticity: Where ecology and evolution meet molecular biology. *Bioessays* 19: 519-525.

Cameron, N. 2002. *Human Growth and Development*. San Diego, Academic Press.

Cardillo, M. 2010. Some applications of geometric morphometrics to archaeology. In Elewa, A. M. T. (ed.); *Morphometrics for Nonmorphometricians*: 325-341. Berlin, Springer-Verlag.

Carroll, S.B., J. K. Greiner, and S. D. Weatherbee. 2001. *From DNA to Diversity*. Victoria, Blackwell Science.

Clarke, D. 1968. *Analytical Archaeology*. London, Methuen.

Clarke, G. M. 1998. The genetic basis of developmental stability. V. Inter- and intra-individual character variation. *Heredity* 80: 562-567.

Clarkson, C. 2002. An index of invasiveness for the measurement of unifacial and bifacial retouch: A theoretical, experimental and archaeological verification. *Journal of Archaeological Science* 1: 65-75.

Clarkson, C. 2005. Tenuous types: Scraper reduction continuums in the Eastern Victoria River Region, Northern Territory. In Clarkson, C. and L. Lamb (eds.); *Lithics 'Down Under': Australian Approaches to Lithic Reduction, Use and Classification*: 21-34. British Archaeological Reports International Monograph Series S1408. Oxford, Archaeopress.

Clarkson, C. 2007. *Lithics in the Land of the Lightning Brothers: The Arhaeology of Wardaman Country, Northern Territory.* Terra Australis 25. Canberra, ANU E Press.

Cleland, C. E. 2002. Methodological and epistemic differences between historical science and experimental science. *Philosophy of Science* 69: 474-496.

Collard, M. and S. Shennan. 2000. Processes of cultural change in prehistory: A case study from the European Neolithic. In Renfrew, C. and K. Boyle (eds.); *Archaeogenetics: DNA and the Population Prehistory of Europe*: 89-97. Cambridge, McDonald Institute for Archaeological Research.

Costa, A. G. 2010. A geometric morphometric assessment of plan shape in bone and stone Acheulean bifaces from the Middle Pleistocene site of Castel di Guido, Latium, Italy. In Lycett, S. J. and P. R. Chauhan (eds.); *New Perspectives on Old Stones: Analytical Approaches to Palaeolithic Technologies*: 23-41. New York, Springer.

Cracraft, J. 1978. Science, philosophy and systematics. *Systematic Zoology* 27: 213-216.

Cracraft, J. 1981. The use of functional and adaptive criteria in phylogenetic systematics. *American Zoologist* 21: 21-36.

Cracraft, J. 2005. Phylogeny and evo-devo: Characters, homology, and the historical analysis of the evolution of development. *Zoology* 108: 345-356.

Darwent, J. and M. J. O'Brien. 2005. Using cladistics to reconstruct lineages of projectile points from northeastern Missouri. In Lipo, C.P., M. J. O'Brien, M. Collard, and S. Shennan (eds.); *Mapping Our Ancestors: Phylogenetic Approaches in Anthropology and Prehistory*: 185-208. New Brunswick, Aldine Transaction.

Davidson, I., 2002. The finished artefact fallacy: Acheulean hand-axes and language origins. In Wray, A. (ed.); *The Transition to Language*: 180-203. Oxford, Oxford University Press.

de Pinna, M. C. C., 1991. Concepts and tests of homology in the cladistic paradigm. *Cladistics* 7: 367-394.

Debat, V. and P. David. 2001. Mapping phenotypes: Canalization, plasticity and developmental stability. *Trends in Ecology & Evolution*, 16: 555-561.

Debat, V., P. Alibert, P. David, E. Paradis and J. C. Auffray. 2000. Independence between developmental stability and canalization in the skull of the house mouse. *Proceedings of the Royal Society of London B* 267: 423-430.

Dibble, H. 1995. Middle Paleolithic scraper reduction: Background, clarification, and review of evidence to date. *Journal of Archaeological Method and Theory* 2: 299-368.

Donoghue, M. J. 1992. Homology. In Keller, E. F. and E. A. Lloyd (eds.); *Keywords in Evolutionary Biology*: 170-179. Cambridge, Mass, Harvard University Press.

Dunnell, R. C. 1970. Seriation method and its evaluation. *American Antiquity* 35: 305-319.

Dunnell, R. C. 1971. *Systematics in Prehistory*. New York, Free Press.

Dunnell, R. C. 1978. Style and function: A fundamental dichotomy. *American Antiquity* 43: 192-202.

Eerkens, J. W., R. L. Bettinger, and R. McElreath. 2005. Cultural transmission, phylogenetics, and the archaeological record. In Lipo, C.P., M. J. O'Brien, M. Collard, and S. Shennan (eds.); *Mapping Our Ancestors: Phylogenetic Approaches in Anthropology and Prehistory*: 169-183. New Brunswick, Aldine Transaction.

Eldredge, N. 1979. Alternative approaches to evolutionary theory. *Bulletin of the Carnegie Museum of Natural History* 13: 7-19.

Eldredge, N. and J. Cracraft. 1980. *Phylogenetic Patterns and the Evolutionary Process*. New York, Columbia University Press.

Ereshefsky, M. 2009. Homology: Integrating phylogeny and development. *Biological Theory* 4: 225-229.

Etges, W. J. 1989. Evolution of developmental homeostasis in *Drosophila mojavensis*. *Evolutionary Ecology* 3: 189-201.

Flenniken, J. J. 1984. The past, present, and future of flintknapping: An anthropological perspective. *Annual Review of Anthropology* 13: 187-274.

Foley, R. A. 1987. Hominid species and stone tool assemblages: How are they related? *Antiquity* 61: 380-392.

Fox, J. G. 2007. *The Mouse in Biomedical Research*. San Diego, Academic Press.

Gaffney, E. S. 1979. An introduction to the logic of phylogenetic reconstruction. In Cracraft, J. and N. Eldredge (eds); *Phylogenetic Analysis and Paleontology*: 79-111. New York, Columbia University Press.

Galton, F. 1889. Discussion on Edward B. Tylor's On a Method of Investigating the Development of Institutions, Applied to Laws of Marriage and Descent. *Journal of the Anthropological Institute of Great Britain and Ireland* 18: 270-272.

Ghiselin, M. 2005. Homology as a relation of correspondence between parts of individuals. *Theory in Biosciences* 124: 91-103.

Grant, T. and A. G. Kluge. 2004. Transformation series as an ideographic character concept. *Cladistics* 20: 23-31.

Griffiths, P. E. 1999. Squaring the circle: Natural kinds with historical essences. In Wilson, R. A. (ed.); *Species: New Interdisciplinary Essays*: 208-228. Cambridge, Mass., The MIT Press.

Griffiths, P. E. 2006. Function, homology, and character individuation. *Philosophy of Science* 73: 1-25.

Griffiths, P. E. 2007. The phenomena of homology. *Biology and Philosophy* 22: 643-658.

Hall, B. K. (ed.). 1994. *Homology: The Hierarchical Basis of Comparative Biology*. San Diego, Academic Press.

Hall, B. K. 2003. Descent with modification: The unity underlying homology and homoplasy as seen through an analysis of development and evolution. *Biological Review* 78: 409-433.

Hall, B. K. 2007. Homoplasy and homology: Dichotomy or continuum? *Journal of Human Evolution* 52: 473-479.

Haszprunar, G. 1992. The types of homology and their significance for evolutionary biology and systematics. *Journal of Evolutionary Biology* 5: 13-24.

Hodder, I. 2001. *Archaeological Theory Today*. Cambridge, Cambridge University Press.

Jablonka, E. and M. J. Lamb. 2005. *Evolution in Four Dimensions. Genetic, Epigenetic, Behavioral, and Symbolic Variation in the History of Life*. Cambridge Mass., The MIT Press.

Jordan, P., T. Mace. 2005. Tracking culture-historical lineages: Can 'descent with modification' be linked to 'association by descent'? In Lipo, C.P., M. J. O'Brien, M. Collard, and S. Shennan (eds.); *Mapping Our Ancestors: Phylogenetic Approaches in Anthropology and Prehistory*: 149-167. New Brunswick, Aldine Transaction.

Jordan, P. and S. Shennan. 2003. Cultural transmission, language, and basketry traditions amongst the California Indians. *Journal of Anthropological Archaeology* 22: 42-74.

Kearney, M. 2007. Philosophy and phylogenetics. Historical and current connections. In Hull, D. and M. Ruse (eds.); *The Cambridge Companion to the Philosophy of Biology*: 211-231. New York, Cambridge University Press.

Kearney, M. and O. Rieppel. 2006. Rejecting "the given" in systematics. *Cladistics* 22: 369-377.

Kleisner, K. 2007. The formation of the theory of homology in biological sciences. *Acta Biotheoretica* 55: 317-340.

Klingenberg, C. P. 2008. Morphological integration and developmental modularity. *Annual Review of Ecology, Evolution and Systematics* 39: 115-132.

Koerner, S. 2011. Experimental archaeology after simplicity: Implications for reflexivity of insights that a "common world" is not a "given". In Millson, D. C. E. (ed.); *Experimentation and Interpretation: The Use of Experimental Archaeology in the Study of the Past*: 61-95. Oxford, Oxbow Books.

Kroeber, A. L. 1931. Historical reconstruction of culture growths and organic evolution. *American Anthropologist* 33: 149-156.

Kuratani, S. 2009. Modularity, comparative embryology and evo-devo: Developmental dissection of evolving body plans. *Developmental Biology* 332: 61-69.

Lao, T. and Y. Tomoyasu. 2011. Morphological homology vs. developmental homology-a case study using insect wing veins. *Abstracts of the 52nd Annual Drosophila Research Conference*, p. 232. San Diego, The Genetics Society of America.

Laubichler, M. 2000. Homology in development and the development of the homology concept. *American Zoologist* 40: 777-788.

Lauder, G. 1994. Homology, form, and function. In Hall, B. (ed.); *Homology: The Hierarchical Basis of Comparative Biology*: 151-196. San Diego, Academic Press.

Liefting, M., A. A. Hoffmann, and J. Ellers 2009. Plasticity versus environmental canalization: Population differences in thermal responses along a latitudinal gradient in *Drosophila serrata*. *Evolution* 63: 1954-1963.

Love, A. C. 2007. Functional homology and homology of function: Biological concepts and philosophical consequences. *Biology and Philosophy* 22: 691-708.

Lycett, S. J. 2007. Why is there a lack of Mode 3 Levallois technologies in East Asia? A phylogenetic test of the Movius-Schick hypothesis. *Journal of Anthropological Archaeology* 26: 541-575.

Lycett, S. J. 2009a. Quantifying transitions: Morphometric approaches to Palaeolithic variability and technological change. In Camps, M. and P. R. Chauhan (eds.); *Sourcebook of Palaeolithic Transitions: Methods, Theories, and Interpretations*: 79-92. New York, Springer.

Lycett, S. J. 2009b. Are Victoria West cores 'proto-Levallois'? A phylogenetic assessment. *Journal of Human Evolution* 56: 175-191.

Lycett, S. J. 2010. Cultural transmission, genetic models, and lithic variability: Integrative analytical approaches. In Lycett, S. J. and P. R. Chauhan (eds.); *New Perspectives on Old Stones: Analytical Approaches to Palaeolithic Technologies*: 207-234. New York, Springer.

Lycett, S. J. and P. R. Chauhan 2010. Analytical approaches to Palaeolithic technologies: An introduction. In Lycett, S. J. and P. R. Chauhan (eds.); *New Perspectives on Old Stones: Analytical Approaches to Palaeolithic Technologies*: 1-22. New York, Springer.

Lyman, R. L. and M. J. O'Brien. 2005. Seriation and cladistics: The difference between anagenetic and cladogenetic evolution. In Lipo, C.P., M. J. O'Brien, M. Collard, and S. Shennan (eds.); *Mapping Our Ancestors: Phylogenetic Approaches in Anthropology and Prehistory*: 65-88. New Brunswick, Aldine Transaction.

Lyman, R. L., S. Wolverton and M. J. O'Brien. 1998. Seriation, superposition, and interdigitation: A history of Americanist graphic depictions of culture change". *American Antiquity* 63: 239-61.

Matthen, M. 2007. Defining vision: What homology thinking contributes. *Biology and Philosophy* 22: 675-689.

Maturana, H. R. and F. J. Varela. 1987. *The Tree of Knowledge: The Biological Roots of Human Understanding*. Boston, Shambhala Publications Inc.

McLennan, D. A. and D. R. Brooks. 2001. Phylogenetic systematics: Five steps to enlightenment. In: Adrain, J. M., G. D. Edgcombe, and B. S. Lieberman (eds.); *Fossils, Phylogeny and Form: An Analytical Approach*: 7-28. New York, Kluwer Academic/Plenum.

Medina, J. 1996. *The Clock of Ages: Why We Age, How We Age, Winding Back the Clock*. New York, Cambridge University Press.

Moczek, A. P. 2008. On the origins of novelty in development and evolution. *BioEssays* 30: 432-447.

Morrison, D. M. 1994. *Validity in Lithic Debitage Analysis: An Experimental Assessment Comparing Quartzite and Obsidian*. Unpublished Master Dissertation. Simon Fraser University. http//www.ir.lib.sfu.ca/bitstream/1892/6513/1/b14830097.pdf. 03/03/2011.

Müller, G. 2003. Homology: The evolution of morphological organization. In Müller, G. and S. Newman (eds.); *Origination of Organismal Form: Beyond the Gene in Developmental and Evolutionary Biology*: 51-69. Cambridge, Mass, The MIT Press.

Nelson, G. J. 1978. Ontogeny, phylogeny, paleontology and the biogenetic law. *Systematic Zoology* 27: 324-345.

Nelson, G. J. 1989. Cladistics and evolutionary models. *Cladistics* 5: 275-289.
Nelson, G. J. and N. I. Platnick. 1981. *Systematics and Biogeography*. New York, Columbia University Press.
Noonan, H. 2009. Identity. In Zalta, E. N. (ed.) *The Stanford Encyclopedia of Philosophy*. http://plato.stanford.edu/archives/win2009/entries/identity/. 02/24/2011.
O'Brien, M. J. (ed.). 2008. *Cultural Transmission and Archaeology: Issues and Case Studies*. Washington, D.C., Society for American Archaeology Press.
O'Brien, M. J. 2010. The future of Paleolithic studies: A view from the New World. In Lycett, S. J. and P. R. Chauhan (eds.); *New Perspectives on Old Stones: Analytical Approaches to Palaeolithic Technologies*: 311-334. New York, Springer.
O'Brien, M. J. and R. L. Lyman. 2000. *Applying Evolutionary Archaeology: A Systematic Approach*. New York, Kluwer Academic/Plenum.
O'Brien, M. J. and R. L. Lyman. 2003. *Cladistics and Archaeology*. Salt Lake City, University of Utah Press.
O'Brien, M. J., J. Darwent, and R. L. Lyman. 2001. Cladistics is useful for reconstructing archaeological phylogenies: Paleoindian points from the southeastern United States. *Journal of Archaeological Science* 28: 1115-1136.
O'Brien, M. J., R. L. Lyman, and J. Darwent. 2002. Cladistics and archaeological phylogeny. In Martínez, G. and J. L. Lanata (eds.); *Perspectivas Integradoras entre Arqueología y Evolución. Teoría, Métodos y Casos de Aplicación*: 175-186. Olavarría, INCUAPA-UNCPBA.
O'Higgins, P. 2000. The study of morphological variation in the hominid fossil record: Biology, landmarks and geometry. *Journal of Anatomy* 197: 103-120.
Owen, R. 1843. *Lectures on the Comparative Anatomy and Physiology of the Invertebrate Animals*. London, Longman, Brown, Green and Longmans.
Patterson, C. 1982. Morphological characters and homology. In Joysey, K. and A. Friday (eds.); *Problems of Phylogenetic Reconstruction*: 21-74. London, Academic Press.
Patterson, C. 1988. Homology in classical and molecular biology. *Molecular Biology and Evolution* 5: 603-625.
Platnick, N. I. 1979. Gaps and prediction in classification. *Systematic Zoology* 27: 472-474.
Platnick, N. I. and H. D. Cameron. 1977. Cladistic methods in textual, linguistic and phylogenetic analysis. *Systematic Zoology* 26: 380-385.
Puelles, L. and L. Medina. 2002. Field homology as a way to reconcile genetic and developmental variability with adult homology. *Brain Research Bulletin* 57: 243-255.
Raff, R. A. 1996. *The Shape of Life*. Chicago, Chicago University Press.
Riede, F. 2006. Chaîne *opèratoire*, chaîne evolutionaire?. Putting technological sequences in evolutionary context. *Archaeological Review from Cambridge* 21: 50-75.
Riede, F. 2008. Maglemosian memes: Technological ontogeny, craft traditions and the evolution of Northern European barbed points. In O'Brien, M. J. (ed.); *Cultural Transmission and Archaeology: Issues and Case Studies*: 178-189. Washington, D.C., Society for American Archaeology Press.
Riede, F. 2009. Tangled trees: Modelling material culture evolution as host-associate co-speciation. In Shennan, S. J. (ed.); *Pattern and Process in Cultural Evolution*: 85-98. Berkely, University of California Press.
Rieppel, O. 1988. *Fundamentals of Comparative Biology*. Basel, Birkhauser Verlag.
Rieppel, O. 2005. Modules, kinds, and homology. *Journal of Experimental Zoology (Mol Dev Evol)* 304B: 18-27.
Rieppel, O., 2008. Species as a process. Acta Biotheoretica 57: 33-49.
Rieppel, O. and M. Kearney. 2002. Similarity. *Biological Journal of the Linnean Society* 75: 59-82.
Rieppel, O. and M. Kearney. 2007. The poverty of taxonomic characters. *Biology and Philosophy* 22: 95-113.
Rose, M. R. and G. V. Lauder. 1996. *Adaptation*. San Diego, Academic Press.
Roth, V. L. 1984. On homology. *Biological Journal of the Linnean Society* 22:13-29.
Roth, V. L. 1988. The biological basis of homology. In C. J. Humphries (ed.); *Ontogeny and Systematics*: 1-26. New York, Columbia University Press.
Salthe, S. N. 1993. *Development and Evolution: Complexity and Change in Biology*. Cambridge, Mass, The MIT Press.
Salthe, S. N. 2010. Development (and evolution) of the Universe. *Foundations of Science* 15: 357-367.

Salthe, S. N. and R. von Sternberg. 2009. Biological complexity: Beyond the genome. In *Encyclopedia of Life Sciences* (ELS). Chichester, John Wiley & Sons.

Scheinsohn, V. 2009. Cladística y arqueología: Razones para un encuentro. In Oliva, F., N. de Grandis, and J. Rodriguez (eds.); *Arqueología Argentina en los Inicios de un Nuevo Siglo* T2: 617-624. Rosario, Editorial Laborde.

Scheinsohn, V., C. Szumik, S. Leonardt, and F. Rizzo. 2009. Distribución espacial del arte rupestre en el bosque y la estepa del norte de Patagonia. Nuevos resultados. In in Salemme, M., F. Santiago, M. Álvarez, E. Piana, M. Vázquez, and M. E. Mansur (eds.); *Arqueología de Patagonia: Una Mirada desde el Último Confín*: 541-560. Ushuaia, Editorial Utopías.

Schlosser, G. and G. P. Wagner (eds.). 2004. *Modularity in Development and Evolution*. Chicago, Chicago University Press.

Schreider, E. 1960. *La Biométrie*. Paris, Presses Universitaires de France.

Shennan, S. J. and M. Collard. 2005. Investigating processes of cultural evolution on the north coast of New Guinea with multivariate and cladistic analyses. In Mace, R., C. J. Holden, and S. J. Shennan (eds.); *The Evolution of Cultural Diversity: A Phylogenetic Approach*: 133-164. London, UCL Press.

Shott, M. 2003. Reduction sequence and *chaîne opèratoire*. *Lithic Technology* 28: 95-105.

Shott, M. 2008. Darwinian evolutionary theory and lithic analysis. In O'Brien, M. J. (ed.); *Cultural Transmission and Archaeology: Issues and Case Studies*: 146-157. Washington, D.C., Society for American Archaeology Press.

Shott, M. 2010. Stone-tool demography: Reduction distributions in Paleoindian tools. In Lycett, S. J. and P. R. Chauhan (eds.); *New Perspectives on Old Stones: Analytical Approaches to Palaeolithic Technologies*: 275-293. New York, Springer.

Shott, M. and P. Sillitoe. 2005. Use life and curation in New Guinea experimental used flakes. *Journal of Archaeological Science* 32: 653-663.

Sillitoe, P. and K. Hardy. 2003. Living lithics: Ethnoarchaeology in Highland Papua New Guinea. *Antiquity* 77: 555-566.

Sluys, R. 1996. The notion of homology in current comparative biology. *Journal of Zoological Systematics and Evolutionary Research* 34: 145-152.

Stearns, S. C., M. Kaiser, and T. J. Kawecki. 1995. The differential genetic and environmental canalization of fitness components in *Drosophila melanogaster*. *Journal of Evolutionary Biology* 8: 539-557.

Stevens, P. F. 1984. Homology and phylogeny: Morphology and systematics. *Systematic Botany* 9: 395-409.

Strauss, R. E. 2010. Discriminating groups of organisms. In Elewa, A. M. T. (ed.); *Morphometrics for Nonmorphometricians*: 73-99. Berlin, Springer-Verlag.

Szalay, F. S. 1977. Phylogenetic relationships and a classification of the eutherian mammalia. In Hecht, M. K., P. C. Goody and B. M. Hecht (eds); *Major Patterns in Vertebrate Evolution*: 315-374. New York, Plenum Press.

Tautz, D. 1998. Debatable homologies. *Nature* 395:17-19.

Tehrani, J. and M. Collard. 2002. Investigating cultural evolution through biological phylogenetic analyses of Turkmen textiles. *Journal of Anthropological Archaeology* 21: 443-463.

Teltser, P. 1995. Culture history, evolutionary theory, and frequency seriation. In Teltser, P. (ed.); *Evolutionary Archaeology: Methodological Issues*: 51-68. Tucson, University of Arizona Press.

Van Dongen, S., and L. Lens 2000. The evolutionary potential of developmental instability. *Journal of Evolutionary Biology* 13: 326-335.

Waddington, C. H. 1957. *The Strategy of the Genes*. London, George Allen & Unwin.

Wagner, G. 1989a. The biological homology concept. *Annual Review of Ecology and Systematics* 20:51-69.

Wagner, G. R. 1989b. The origin of morphological characters and the biological basis of homology. *Evolution* 43: 1157-1171.

Wagner, G. R. 1996. Homologues, natural kinds and the evolution of modularity. *American Zoologist* 36: 36-43.

Wagner, G. R. 1999. A research programme for testing the biological homology concept. In Bock, G. R. and G. Cardew (eds.); *Homology*: 125-134. Novartis Foundation for Gerontological Research Symposium 222. Chichester, John Wiley & Sons.

Wagner, G. R. 2007. The developmental genetics of homology. *Nature Reviews Genetics* 8: 473-479.

Wagner, G. R., M. Pavlicev, and J. M. Cheverud. 2007. The road to modularity. *Nature Reviews Genetics* 8: 921- 931.

Wake, D. 2003. Homology and homoplasy. In Hall, B. K. and W. M. Olson (eds.); *Keywords and Concepts in Evolutionary Developmental Biology*: 191-200. Cambridge Mass., Harvard University Press.

Wenzel, J. W. 1992. Behavioral homology and phylogeny. *Annual Review of Ecology and Systematics* 23: 361-381.

Whittaker, J. 1994 *Flintknapping: Making and Understanding Stone Tools.* Austin, University of Texas Press.

Wiley, E. O. 1975. Karl R. Popper, systematics, and classification: A reply to Walter Bock and other evolutionary taxonomists. Systematic Zoology 24: 233-243.

Wiley, E. O. 1976. The phylogeny and biogeography of fossil and recent gars (Actinopterygii: Lepisosteidae). *Miscellaneous Publications of the Museum of Natural History of the University of Kansas* 64: 1-111.

Williams, D. M., M. C. Ebach, and G. Nelson 2008. *Foundations of Systematics and Biogeography.* New York, Springer.

Wilson, J. K. and W. Andrefsky, Jr. 2008. Unpacking production, resharpening and hammer type, in Andrefsky, Jr., W. (ed.); *Lithic Technology: Measures of Production, Use and Curation*: 86-105. Cambridge, Cambridge University Press.

Winther, R. G. 2009. Character analysis in cladistics: Abstraction, reification, and the search for objectivity. *Acta Biotheoretica* 57: 129-162.

HUMAN HOLOCENE COLONIZATION, DIET BREADTH AND NICHE CONSTRUCTION IN SIERRAS OF CORDOBA (ARGENTINA)

Diego Rivero[1] and Matías Medina[2]

[1,2]CONICET - Área de Arqueología y Etnohistoria, Centro de Estudios Históricos "Prof. Carlos S. A. Segreti", M. C. del Corro 308, Córdoba, Argentina. E-mail: ayampitin1@yahoo.com.ar, paleomedina@yahoo.com.ar

Abstract

This chapter focuses on how human colonization of Sierras of Cordoba (Argentina) impact on the high-ranked prey resource density *–i.e. Lama guanicoe-*. Three hypothetical scenarios were modelled altering the encounter rate of *L. guanicoe* and predicting how its variation should broaden or narrow the optimal diet of human societies. Diachronic trends in the *Lama* sp. encounter rate were analyzed with two quantitative indeces that compare its relative abundance in the assemblages: the "camelid index" and the "artiodactyls index". The archaeofaunal record reflects the temporal decrease in Lama sp. frequency through the Holocene. This taxonomical trend is associated to a decrease in their encounter rate as a result of human over-explotation under specific ecological conditions. The results draw to the conclusion that a broad-spectrum diet which may not be an exclusively Late Holocene phenomenon and its origins may be found in earlier processes beginning during the Middle Holocene. Thus, the incorporation of small-scale agriculture around ca. 1200-1000 BP cannot be satisfactorily explained only by external factors such as climate changes, the diffusion and/or arrival of new biological populations. Research should consider that past human cultural activities played a dynamic role in this process through niche construction, co-directing subsequent cultural changes that are seen in the archaeological record.

KEYWORDS: ARCHAEOLOGY, NICHE CONSTRUCTION, SIERRAS DE CÓRDOBA

Resumen

Este capítulo se focaliza en como la colonización humana de las Sierras de Córdoba (Argentina) impactó sobre la densidad del recurso del mayor ranking –i.e. *Lama guanicoe-*. Se plantean 3 escenarios hipotéticos alterando la frecuencia de encuentro con *L. guanicoe* y prediciendo como su variación va a ampliar o estrechar la dieta óptima de las sociedades humanas. Las tendencias diacrónicas en las frecuencias de encuentro con *L. guanicoe* se analizaron con 2 índices cuantitativos que comparan su abundancia relativa en los conjuntos: el "índice de camélidos" y el "índice de artiodáctilos". El registro arqueofaunístico refleja una caída en la frecuencia de *Lama* sp. a través del Holoceno. Este patrón es asociado a una reducción en la tasa de encuentro como resultado de la sobreexplotación humana bajo condiciones ecológicas específicas. Los resultados indican que una dieta de amplio espectro no es un fenómeno exclusivamente del Holoceno Tardío y sus orígenes deben buscarse tempranamente en el Holoceno Medio. En consecuencia, la incorporación de prácticas agrícolas a pequeña escala ca. 1200-1000 AP no puede ser explicada satisfactoriamente por factores externos, como cambios climáticos, difusión y/o el arribo de nuevas poblaciones biológicas. Las investigaciones deben considerar que las actividades culturales prehispánicas jugaron un rol dinámico en este proceso a través de la construcción de nicho, codirigiendo los subsecuentes cambios culturales que se observan en el registro arqueológico.

PALABRAS CLAVE: ARQUEOLOGÍA, CONSTRUCCIÓN DE NICHO, SIERRAS DE CÓRDOBA

Introduction

Humans, as many organisms, modify local environments through resource selection, emiting detritus, constructing shelters and pathways, etc. with spectacular ecological and evolutionary consequences. This process is known in Evolutionary Biology as "niche engineering" or more recently "niche construction" and consider that no species has more greatly amplified this ability than humans (Laland et al 2001; Odling-Smee 2007; Odling-Smee et al 2003; B. Smith 2007). Thus, *Homo sapiens* not only adapts to different environmental conditions but also interacts with them, modifying natural selective pressures which will be inherited to his descendants (Laland et al 2001).

Human Holocene colonization of Sierras of Cordoba (Argentina) and its impact on the high-ranked prey resource density –e.g. *Lama guanicoe*– are analyzed under this perspective. Prey density simulations and Diet Breadth Model predictions (Stephen & Krebs 1986) were used to assess human predation pressure on local *L. guanicoe* demography. Niche Construction Theory (Odling-Smee et al 2003) is used as a behavioral and evolutionary framework to understand main changes in Sierras of Cordoba Holocene archaeological record, including the increased importance of low-return resources and the incorporation of small-scale agriculture at the end of the Holocene.

This chapter is focused on faunal analysis from six archaeological sites located in Sierras of Cordoba upper mountain grassland range dated on from Early Holocene to Late Holocene. This environment was the habitat for ungulate herds such as the guanaco (*Lama guanicoe*) and pampas deer (*Ozotoceros bezoarticus*), main profitable prey for Holocene hunter-gatherers and horticulturalists. Thus, temporal variations in taxonomic resource abundance provide an excellent opportunity to examine the resource intensification process previously proposed for the region (e.g. Berberián et al 2008; Medina 2008; Medina & Rivero 2007; Pastor 2007a; Recalde 2009; Rivero 2009;).

Environment and Human Colonization of the Sierras of Cordoba

The Sierras of Cordoba are a low-altitude mountain range (500-2790 m) with a complex of peaks, valleys, steep slopes and high-plains that are especially interesting from a human evolutionary ecology point of view. In a simplified scheme, most of the regional plant communities were included in the Sierra Chaco which is a semi-arid xerofic forest dominated by small trees and shrubs with edible wild fruits such as algarrobo (*Prosopis* spp.), chañar (*Geoffroea decorticans*), mistol (*Zizyphus mistol*), molle de beber (*Lithraea molleoides*), etc. (Luti *et al.* 1979; Demaio *et al.* 2002) (Figure 1). The fauna of Sierra Chaco is dominated by low-diversity neotropical small-sized mammals adapted to closed-vegetation environments such as the brown-brocket deer (*Mazama guazoupira*), collared peccary (*Pecari tajacu*), vizcachas (*Lagostomus maximus*), armadillos (Dasypodidae spp.), small caviomorph rodents (Caviinae spp. and Ctenomyinae spp.), etc. (Bucher and Abalos 1979).

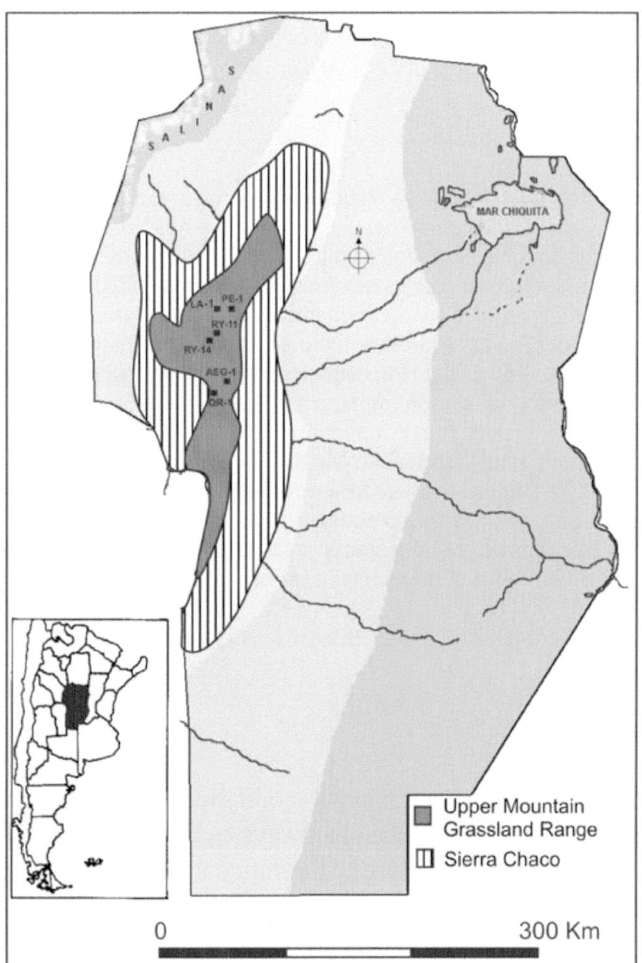

FIGURE 1. MAP OF CÓRDOBA PROVINCE (ARGENTINA) INDICATING SITE LOCATIONS DISCUSSED IN THE TEXT.

Above 1500 m changes in floristic composition though altitudinal gradient results in high-plain known as upper mountain grassland range (Luti *et al.* 1979). From South to North the main high-plains are Pampa de Achala (2,000 m), Pampa de San Luís (1,900 m) and Pampa de Olaen (1,100 m). Zoogeographically they are included in the Sub-Andean District of the Andean-Patagonian Sub-region (Ringuelet 1961) which included now-extincted prey species of great economic importance due to their body size, such as guanaco, pampas deer, taruca (*Hippocamelus* sp.) and the south-american ostrich (*Rhea* sp.). Thus, the upper mountain grassland range constitutes a "faunistic island" of 6,000 km^2 surrounded by chaquenian environments, providing geographical isolation and refuge to an interesting Andean-Patagonian biota (Figure 1) (Bucher & Abalos 1979; Cabido *et al.* 1998; Cei 1972; Polop 1989; Ringuelet 1961).

Palaeoclimatic information indicates that the mountain vegetation during the Pleistocene-Holocene Transition was different from the present. The grassland has a greater extension due to cold and dry climate conditions, connecting the high-plain to lowland grasslands, precordilleran and cordilleran zones of central and northwestern Argentina (Adams & Faure 1997; Carignano 1999; Piovano et al 2010; Sanabria & Argüello 2003). This allowed the flow of several Andean-Patagonian faunal species, including *L. guanicoe*. The establishment of subtropical conditions at 6000 BP redefined the landscape with the formation of the Sierra Chaco vegetational belt and the biogeographical isolation of Andean-Patagonian biota above 1,500 m (Sanabria & Argüello 2003; Teta *et al.* 2005).

There is a combination of factors that enhance the chance that the "faunistic island" of the upper mountain grassland range will be highly-vulnerable to human predation (Nagaoka 2002:422). First, population size is generally smaller and the loss of individuals within this population is often more severe. Second, geographic isolation means that there is no replacement population, reducing the genetical flow and decreasing its long-term reproductive viability. Finally, preys will be severely impacted by the arrival of new predators in the local ecosystem, mainly if the predators were efficient hunters such as humans.

The earliest evidence of humans in the Sierras of Cordoba is dated at ca. 11,000 BP and it was obtained at the El Alto 3 site, on the Pampa de Achala (Rivero & Roldán 2005). However, the human occupation during the Pleistocene-Holocene Transition was not intense, with a low-visibility archaeological record. After 8,000 BP the regional archaeological record increases progressively its visibility, reaching its highest level towards the end of the Late Holocene (Berberián et al 2008; Rivero 2009; Rivero & Berberián 2008). During this time the Sierras of Cordoba evidenced a process of colonization and stabilization of human occupation (sensu Borrero 1994-1995), involving population growth, the seasonal use of low-ranked mountain microenvironments and the incorporation of small-scale agriculture ca. 1200-1000 BP (Berberián et al 2008; Pastor 2006; Medina 2008).

The high-plains were exploited as the main large-game hunting territory through the Holocene sequence, so the isolated artiodactyl herds were subjected to a continuous human predation pressure during this time. This situation may cause anthropogenic-induced reduction in ungulate prey abundance so that human foragers should respond by increasing diet breadth (Medina & Rivero 2007). Thus, the archaeofaunal record available for the Sierras of Cordoba upper mountain grassland range ca. 7200-360 BP must be evaluated to assess the human impact on the prehistoric environment and how foragers adapt to the changing environmental situations that they create. This procedure would also offer valuable information about past ecosystems and how they changed during the Holocene, as well as measure human contribution to that process (Grayson 2001; Lyman & Cannon 2004).

Evolutionary Ecology, Niche Construction and Diet Breadth Model

Evolutionary Ecology (sometimes referred to as Human Behavioral Ecology) is the study of adaptative design and its ecological setting through the application of natural selection theory (Winterhalder & Smith 1992; Bird & O´Connell 2006; Broughton et al 2010). Human capacity for rapid adaptative adjustments to changing socio-ecological conditions is considered the product of natural selection, whereas specific behavioral response are not (Boone & Smith 1998; O´Connell 1995; Winterhalder & Smith 1992). The Evolutionary Ecology approach uses formal optimization models to explore the fitness-related cost and benefits of behavioral alternatives in specific socio-ecological context.

Thus, the environment is considered a key variable to explain variability in human behaviors (Bird & O´Connell 2006; Winterhalder & Smith 1992). As human behaviors leaves intencional or unintencional material traces –i.e. the archaeological record– Evolutionary Ecology can generate expectations about prehistoric behaviors and their empirical consequences (Bird & O´Connell 2006; O´Connell 1995).

Because Evolutionary Ecology often focuses on how behavior adjusts to environmental changes extrinsic to the organism as well as those resulting from its own activities, this framework is inextricably tied to Niche Construction Theory (Broughton et al 2010). The central premise of the niche construction perspective is that human cultural activities play an active role in the evolutionary process modifying the selective pressures of the environment that drive further behavioral, cultural and genetical changes in human population (Odling-Smee et al 2003). Ecological inheritance is a key concept in this theoretical framework, refering to the selective environments that the new generations inherit from their genetic or ecological ancestors, who have modified it on evolutionary time-scales through niche construction (Laland et al 2001; Odling-Smee 2007; Odling-Smee et al 2003).

A basic model of Evolutionary Ecology easy to integrate with the niche construction approaches is the Diet Breadth Model, because it draws on the same evolutionary logic (Broughton et al 2010). The Diet Breadth Model is specifically designed to predict which resources will be pursued or ignored when encountered by forager searching in a homogeneous environment (Bettinger 1991; Hill & Hawkes 1983; Kaplan & Hill 1992). It considers that the main objective of foraging activities is to maximize the capture of energy or post-encounter return rate –i.e. the amount of energy gathered per unit of time after encountering a resource (*sensu* Kelly 1995)–. Thus, foragers concentrate this effort pursuing a set of high-return prey types rather than taking low-return smaller animals when encountered.

The Diet Breadth Model predicts that prey types were included in the diet according to their profitability. Those with a high post-encounter return rate should always remained in the diet and were pursued upon enconunter. Thus, its archaeological frequency reflects their abundance in the local environment (Broughton & Grayson 1993; Grayson et al 2001). Only if the encounter rate of the high-return prey types decreases due to climate changes and/or human-induced prey depression (Broughton 1999) the diet should be broadened to include more abundant low-return resources (Bettinger 1991; Bird & O´Connel 2006; Broughton 1994; Kaplan & Hill 1992; Kelly 1995; E. Smith 1983; Winterhalder 1981a, 1981b; Winterhalder & Smith 2000). However, prey types with lower profitability than the overall net caloric return rate should always be ignored regardless of their abundance in the landscape. This hypothetical point is defined as the "Optimal Diet Breadth" or "Local Optimum" (Kaplan & Hill 1992; E. Smith 1983; Winterhalder 1981a). This is:

$$E_j / h_j > \sum \lambda_i * E_j / \sum \lambda_i * h_i + 1 , \qquad (1)$$

where λ_i is the encounter rate of each resource, Ej the net energy obtained and hj the handling time (Muscio 2004).

The application of the Diet Breadth Model in archaeology requires that the potential preys were ordinal-ranked according to their profitability. Thus, a ranking was built based on the post-encounter return rate of the main taxa available in the Sierras of Cordoba upper mountain grassland range (Table 1). Caloric yields and handling times were obtained from similar taxa from the taxonomic, anatomical and metrical points of view (Hill & Hawkes 1983; Medina 2008; Muscio 1999, 2004; Yacobaccio *et al.* 1998). Only 60% of artiodactyls' eviscerated carcass was considered edible, while 70% of bird and 85% of small mammal carcasses were suitable for food (Byers & Ugan 2005) (Table 1).

Common name	Body size (kg)	Calories/kg	Caloric yield	Handling time (hr)*	Density (ind./km^2)	References
Guanaco	140	1090	91560	3,3	10	Parera (2002)
Greater Rhea	30	1900	39900	1,46	6,7	Medina (2008)
Taruca	70	1200	50400	2,5	1	Parera (2002)
Pampas Deer	45	1200	32400	1,67	5	Parera (2002)
Lesser Rhea	20	1900	26600	1,46	5	Medina (2008)
Armadillo	3,6	3000	4832	0,929	60	Parera (2002)
Duck	1	1230	861	0,369	115	Byers y Ugan (2005)
Pampas Vizcacha	8	1141	4792	1,457	700	Parera (2002)
Tinamou	0,34	1900	452,2	0,37	7,5	Medina(2008)
Tuco-tuco	0,7	1200	714	0,35	1271	Parera (2002)
Cuy	0,25	1200	252	0,35	2400	Parera (2002)

*: Handling times refers to the post-encounter activities incumbent on the depredator to transfer a living prey item from the environment into its digestive tract, including pursuing, butchering, processing and consuming resources. These data was obtained from Medina (2008).

TABLE 1. PARAMETERS CONSIDERED IN THE OPTIMAL DIET BREADTH.

L. guanicoe is the high-ranked prey type followed by the south-american ostrich, cervids and, finally, small rodents.

According to Diet Breadth Model, *Lama* sp. will be taken whenever possible and its archaeological frequency should reflect its abundance during the Holocene. This reasoning allows three hypothetical scenarios to be modelled altering the encounter rate of *L. guanicoe* and predicting how its variation should broaden or narrow the optimal diet of human societies.

Scenario 1:

Considering densities of 10 ind./km^2 for *L. guanicoe*, the optimal diet includes artiodactyls that inhabited the Sierras of Cordoba whose body size varies from ca. 30 kg (*O. bezoarticus*) to ca. 140 kg (*L. guanicoe*). Flightless-birds *Rhea* sp. with a body size of 20-30 kg are also be exploited. Instead, small and medium-sized mammals and other vertebrates weighting less than 15 kg will be dropped from the diet (Table 2).

Common name	λ_i Encounte r rate (ind./hr)	h_j Handling times (hr)	E_j Kilocaloric yield	E_j/h_j Post-encounter return rate (Kcal/hr)	$\lambda_i \cdot E_j$ Encounter rate / Kcal	$\lambda_i \cdot h_j$ Handling time (hr/Kg)	? $\lambda_i \cdot h_i + 1$ Pursuing + handling times (hr)	? $\lambda_i . E_j$ (Kcal/hr)	? $\lambda_i \cdot E_j$ / ? $\lambda_i . h_i + 1$ Overall Return rate (Kcal/hr)
Guanaco	0,07	3,3	91560	27745	6409,2	0,231	1,231	6409,2	5206
Greater Rhea	0,046	1,46	39900	27328	1835,4	0,06716	1,29816	8244,6	6350
Taruca	0,007	2,5	50400	20160	352,8	0,0175	1,31566	8597,4	6534
Pampas deer	0,035	1,67	32400	19401	1134	0,05845	1,37411	9731,4	7081
Lesser Rhea	0,035	1,46	26600	18219	931	0,0511	1,42521	10662,4	7481
Armadillo	0,42	0,929	4832	5201	2029,44	0,39018	1,81539	12691,84	6991
Pampas vizcacha	4,9	1,457	4792	3288	23480,8	7,1393	8,95469	36172,64	4039
Duck	0,81	0,369	861	2333	697,41	0,29889	9,25358	36870,05	3984
Tinamou	0,05	0,37	452,2	1222	22,61	0,0185	9,27208	36892,66	3978
Tuco-tuco	8,9	0,35	408	1165	3631,2	3,115	12,38708	40523,86	3271
Cuy	16,8	0,35	252	720	4233,6	5,88	18,26708	44757,46	2450

Note: The shaded area shows the taxa included in the optimal diet breadth.

TABLE 2. OPTIMAL DIET BREADTH ACCORDING TO SCENARIO 1.

Scenario 2:

With a density of 5 ind./km^2 for *L. guanicoe*, the optimal diet shows no significant difference from Scenario 1. Artiodactyls and *Rhea* sp. will be pursued during subsistence activities. Medium and small-vertebrates continue being ignored (Table 3).

Scenario 3:

Finally, *L. guanicoe* densities of 1 ind./km^2 reduce the encounter rate with the high-ranked resource and the model predicts the addition of medium-sized mammals (ca. 5 kg) such as dasypodids (Table 4). Artiodactyls and *Rhea* sp. continue being the main pursued prey due to their profitability, but *L. guanicoe* will be found less frequently in deposits. Smaller-sized taxas are still not included in the diet.

Common name	λ_i Encounter rate (ind./hr)	h_i Handling times (hr)	E_i Kilocaloric yield	E_i / h_i Post-encounter return rate (Kcal/hr)	$\lambda_i \cdot E_i$ Encounter rate / Kcal	$\lambda_i \cdot h_i$ Handling time (hr/Kg)	? $\lambda_i \cdot h_i + 1$ Pursuing + handling times (hr)	? $\lambda_i . E_i$ (Kcal/hr)	? $\lambda_i \cdot E_i$ / ? $\lambda_i . h_i + 1$ Overall Return rate (Kcal/hr)
Guanaco	0,035	3,3	91560	27745	3204,6	0,1155	1,1155	3204,6	2872
Greater Rhea	0,046	1,46	39900	27328	1835,4	0,06716	1,18266	5040	4261
Taruca	0,007	2,5	50400	20160	352,8	0,0175	1,20016	5392,8	4493
Pampas deer	0,035	1,67	32400	19401	1134	0,05845	1,25861	6526,8	5185
Lesser Rhea	0,035	1,46	26600	18219	931	0,0511	1,30971	7457,8	5694
Armadillo	0,42	0,929	4832	5201	2029,44	0,39018	1,69989	9487,24	5581
Pampas vizcacha	4,9	1,457	4792	3288	23480,8	7,1393	8,83919	32968,04	3729
Duck	0,81	0,369	861	2333	697,41	0,29889	9,13808	33665,45	3684
Tinamou	0,05	0,37	452,2	1222	22,61	0,0185	9,15658	33688,06	3679
Tuco-tuco	8,9	0,35	408	1165	3631,2	3,115	12,27158	37319,26	3041
Cuy	16,8	0,35	252	720	4233,6	5,88	18,15158	41552,86	2289

Note: The shaded area shows the taxa included in the optimal diet breadth.

TABLE 3. OPTIMAL DIET BREADTH ACCORDING TO SCENARIO 2.

Common name	λ_i Encounter rate (ind./hr)	h_i Handling times (hr)	E_i Kilocaloric yield	E_i / h_i Post-encounter return rate (Kcal/hr)	$\lambda_i \cdot E_i$ Encounter rate / Kcal	$\lambda_i \cdot h_i$ Handling time (hr/Kg)	? $\lambda_i \cdot h_i + 1$ Pursuing + handling times (hr)	? $\lambda_i . E_i$ (Kcal/hr)	? $\lambda_i \cdot E_i$ / ? $\lambda_i . h_i + 1$ Overall Return rate (Kcal/hr)
Guanaco	0,007	3,3	91560	27745	640,92	0,0231	1,0231	640,92	626
Greater Rhea	0,046	1,46	39900	27328	1835,4	0,06716	1,09026	2476,32	2271
Taruca	0,007	2,5	50400	20160	352,8	0,0175	1,10776	2829,12	2553
Pampas deer	0,035	1,67	32400	19401	1134	0,05845	1,16621	3963,12	3398
Lesser Rhea	0,035	1,46	26600	18219	931	0,0511	1,21731	4894,12	4020
Armadillo	0,42	0,929	4832	5201	2029,44	0,39018	1,60749	6923,56	4307
Pampas vizcacha	4,9	1,457	4792	3288	23480,8	7,1393	8,74679	30404,36	3476
Ducks	0,81	0,369	861	2333	697,41	0,29889	9,04568	31101,77	3438
Tinamou	0,05	0,37	452,2	1222	22,61	0,0185	9,06418	31124,38	3433
Tuco-tuco	8,9	0,35	408	1165	3631,2	3,115	12,17918	34755,58	2853
Cuy	16,8	0,35	252	720	4233,6	5,88	18,05918	38989,18	2158

Note: The shaded area shows the taxa included in the optimal diet breadth.

TABLE 4. OPTIMAL DIET BREADTH ACCORDING TO SCENARIO 3.

Sites, faunal record and methodology

The study focused on 15528 bone specimens collected in the archaeological sites Arroyo El Gaucho 1, Quebrada del Real 1, Rio Yuspe 11, Rio Yuspe 14, Los Algarrobos 1 and Puesto La Esquina 1 (Figure 1, Table 5). During its excavations, sediments were sieved with a fine-screen 2 mm mesh to allow recovery of small vertebrate and macro-botanical remains. The archaeofaunal data was ordered by radiocarbon dating (Medina 2008; Rivero 2009; Rivero *et al.* 2010).

Arroyo El Gaucho 1 (AEG-1; 1843 m) is a rock-shelter situated in Quebrada del Condorito National Park in the Pampa de Achala (Rivero 2009). Two overlapped cultural Components were detected in stratigraphy during its excavation. Lower component (C1) is associated with lanceolate projectile points and bifaces that were radiocarbon dated at 7160±90 years BP (LP-1722, charcoal). The upper component (C2) is characterized by triangular projectile points with slightly to markedly concave bases radiocarbon dated at 3590±60 years BP (LP-1599; charcoal) and 3700±70 years BP (LP-1612; charcoal).

Taxón	AEG1 (C1)	QR1 (C1)	AEG1 (C2)	QR1 (C2)	RY11	LAI1	RY14	PE1
Lama cf. *L. guanicoe*	649	100	355	277	141	4	7	57
Ozotoceros bezoarticus	11	0	27	0	2	1	0	238
cf. *Hippocamelus*	0	1	0	1	0	0	0	0
Odocoleinae	0	0	0	6	22	0	4	0
Artiodactyla	288	25	61	24	71	20	5	82
Lagostomus maximus	0	0	0	0	0	0	1	1
Microcavia australis	0	9	0	59	0	0	0	0
Galea leucoblephara	0	103	0	393	0	0	0	0
Caviinae	316	243	88	859	33	0	2	47
Ctenomys sp.	2	264	7	587	0	5	0	8
Caviiomorpha	0	0	0	1	0	0	0	4
Oxymycterus	0	0	0	0	0	0	0	1
Reithrodon sp.	0	0	0	2	0	0	0	0
Holochilus sp.	3	8	3	44	0	0	0	8
Cricetidae	0	0	0	6	6	0	0	5
Rodentia	5	6	317	25	6	3	1	11
Dasipodidae	3	0	3	0	22	4	3	36
Pseudalopex sp.	0	1	0	10	0	0	0	0
Felinae	0	0	0	1	0	0	0	0
Carnivora	0	0	0	11	0	0	0	0
Mammalia (med.-large size)	1628	139	1065	236	598	67	78	199
Mammalia (small size)	35	5	16	4	17	1	3	7
Anura cf. Leptodactylidae	0	0	0	0	0	0	0	4
Tupinambis sp.	0	0	0	0	0	0	0	20
Sauria sp.	0	0	0	0	0	0	0	1
Rheidae (eggshell)	0	0	0	0	9	0	5	33
Tupinambis sp.	0	0	0	0	0	0	0	20
Sauria sp.	0	0	0	0	0	0	0	1
Rheidae (eggshell)	0	0	0	0	9	0	5	33
Fulica sp.	0	0	0	3	0	0	0	0
Eudromia sp.	0	2	0	0	0	0	0	3
Nothura sp.	0	0	0	0	0	0	0	28
Tinamidae	0	0	0	1	2	0	0	2
Anatidae	1	0	0	1	0	0	0	0
Passeriformes	0	0	0	1	0	0	0	0
Falconiformes	0	2	0	1	0	0	0	0
Aves (small size)	4	0	2	7	0	1	1	7
NUSP	589	1	298	237	1853	302	64	1910
NSP	3534	909	2242	2797	2752	408	174	2712

From Medina (2008, 2009), Pastor (2005, 2007b) y Rivero et al (2010)

Table 5. Number of Identified Specimens per Taxon (NISP) for the upper mountain grassland range assemblages (Córdoba Province, Argentina).

Quebrada del Real 1 (QR1; 1914 m) is a rock-cave located in the Pampa de Achala Provincial Hydrical Reserve (Rivero *et al.* 2010). Three cultural Components were detected during it excavation. The lowest component (C1) was associated to lanceolate projectile points and charcoal dated at 5980±50 years BP (LP-2133). The middle component (C2) is characterized by the presence of triangular projectile points with slightly to markedly concave bases dated at 2950±90 years BP (LP-2042; charcoal). Finally, the top component (C3) contain small-stemmed triangular arrowpoints as well as pottery fragments. This assemblage was temporal-assigned to the Late Holocene but not included in this paper due to absence of radiocarbon dates.

Río Yuspe 11 (RY11; 1810 m) is a rock-shelter located in Pampa de Achala (Pastor 2007a, 2007b). Due to the high-frequency of surface unmovable grinding artifacts and the stratigraphical data, the occupation was associated to animal and vegetal large-group food processing and consumption tasks. The site was dated at 1540±50 BP (LP-1658; charcoal) and 1170±50 BP (LP-1449; charcoal). This chronology positions RY11 in a contemporary context to the spread of some technological innovation such as pottery and small-scale agriculture.

Río Yuspe 14 (RY14; 1860 m) is a small rock-shelter with archaeological occupations dated at 640±70 BP (LP-1514; charcoal). The archaeological remains indicate domestic activities, possibly by small-family units related to Late Holocene seasonal large-game hunting dispersal of co-residential groups from Sierra Chaco environment (Pastor 2005, 2007a).

Los Algarrobos 1 (LAl1, 1100 m a.s.l) is an open-air archaeological site located in the Pampa de Olaen (Medina 2008). The analysis of superficial and stratigraphical archeological remains confirmed its use as an agricultural and residential site dated at 949±40 BP. (AA64818, charcoal).

Finally, Puesto La Esquina 1 (PE1, 1140 m) is an open-air site located in the associated-rich soil of a well-protected gorge of the Pampa de Olaen (Medina 2008). The pottery, bone, lithic and palaeobotanical evidence, as well as an archaeological floor and a hearth, indicate the development of multiple-activities such as the production, storage, processing, cooking, and consumption of foodstuffs, the manufacture, maintenance, and discarding of artifacts, etc.. Thus, PE1 was interpreted as a semi-sedentary residential base or village. Two charcoal samples from overlapping layers were dated at 365±38 BP. (AA64816) and 362±43 BP. (AA64815), placing its occuppations near to the Spaniard arrival and indicating the high-resolution of the deposit.

The high-frequency of burning damage, cutmarks and percussion marks on bones, not treated in depth in this paper, indicate that humans were the main agents that formed these faunal assemblages (Medina 2008; 2009; Medina & Rivero 2007; Medina et al 2009; Rivero 2009; Rivero et al 2010; Rivero et al 2010; Pastor 2005, 2007a; 2007b). Thus, the samples offer an excellent opportunity to assess without taxonomic ambiguity changes in human diet breadth through the Holocene.

Taxonomic abundance was quantified as the Number of Identified Specimens per Taxon (NISP). Teeth were quantified as separate entities, regardless of whether they are found articulated or isolated. This methodology facilitates the interassemblage comparability despite post-depositional processes that might have affected them (Lyman 2008; Stiner 1994). Number of Specimens (NSP) and Number of Unidentified Specimens (NUSP) were also considered (Lyman 2008).

Diachronic trends in the *Lama* sp. encounter rate were analysed with two quantitative indeces that compare its relative abundance in the assemblages at a different taxonomical resolution. The "camelid index" shows the relation between Camelidae and Cervidae bone specimens calculated as:

$$Ic = \frac{\sum NISP\ Lama\ \text{sp.}}{\sum (NISP\ Lama\ \text{sp.} + NISP\ \text{Cervidae})} \quad (2)$$

Instead, the "artiodactyl index" reflects the relation between the artiodactyls and small-vertebrates bones frequency with the same logic that underlies the previous index:

$$Ia = \frac{\sum NISP\ (Lama\ \text{sp.} + \text{Cervidae} + \text{Artiodactyls})}{\sum NISP\ (Lama\ \text{sp.} + \text{Cervidae} + \text{Artiodactyls} + \text{small-vertebrate})} \quad (3)$$

Results and Discussion

The model predicts that artiodactyls and *Rhea* sp. should always be in the optimal diet breadth. Only when the density of guanaco decreases to 1 ind./km^2, armadillos would be an economical option in the evolutionary ecology optimization point of view. Other small-vertebrates such as caviomorph-rodents remained theorically excluded from the diet, even when *L. guanicoe* has a low-density values.

The upper mountain grassland range zooarchaeological record is in agreement with these expectations (Table 5). Indeed, the artiodactyls were exploited across the Holocene (Figure 2). Camelids were the main prey-type exploited by human foragers since the Early Holocene (Rivero 2009). However, *Lama* sp. decreases its archaeological abundance, particularly towards the Late Holocene (Table 6 & 7, Figures 2, 3, & 4).

Conversely, cervids such as *Ozotoceros* sp. increase their frequency since the Middle Holocene and they are the most abundant prey in Late Holocene assemblage of Puesto La Esquina 1 (Medina 2009; Medina & Rivero 2007). However, the prehispanic *Ozotoceros* sp. bone remains look like larger than present-day population parameters, stressing the possibility of a extinct local subspecies, species or even a new genus (see González 1960; Menghin & González 1954). Future studies of metrical and non-metrical comparative variation as well as DNA molecular analysis should solve this taxonomical question.

Few bone specimens of Odocoileinae cf. *Hippocamelus* sp. were identified in Early and Middle Holocene assemblages of Quebrada del Real 1. *Hippocamelus* sp. was also found in Middle Holocene deposits of the classic archaeological sites Ongamira and Intihuasi caves (González 1960; Menghín & González 1954). If the geographical distribution of *Hippocamelus* sp. in the Sierras of Cordoba was confirmed, this would be associated with low-density values and its absence in Late Holocene deposits would be explained by its local extinction.

The medium-sized mammal Euphractinae spp. were also identified in the assemblages. Their archaeological frequency is low in the Early Holocene deposits, but bony dermal scutes of these armadillos increase towards the Late Holocene (see Medina 2009). According to model predictions, this taxonomical pattern indicates a decrease in the *L. guanicoe* encounter rate. Thus, foragers would have to concentrate their subsistence effort pursuing lower-ranked prey than artiodactyls.

The predation of the south-american ostrich is only suggested by low-quantities of eggshell specimens, which are temporally-restricted to Late Holocene deposits. Rhea bones were not found. A possible explanation of its absence may be related to the fact that carcasses were not discarded in the sites, were fragmented into small-unidentified pieces and/or were destroyed by bone mineral density-mediated attritional processes. However, the low-NISP of eggshells might suggest that this flight-less bird did not inhabit the upper mountain grassland range and their eggs were transported from lowlands.

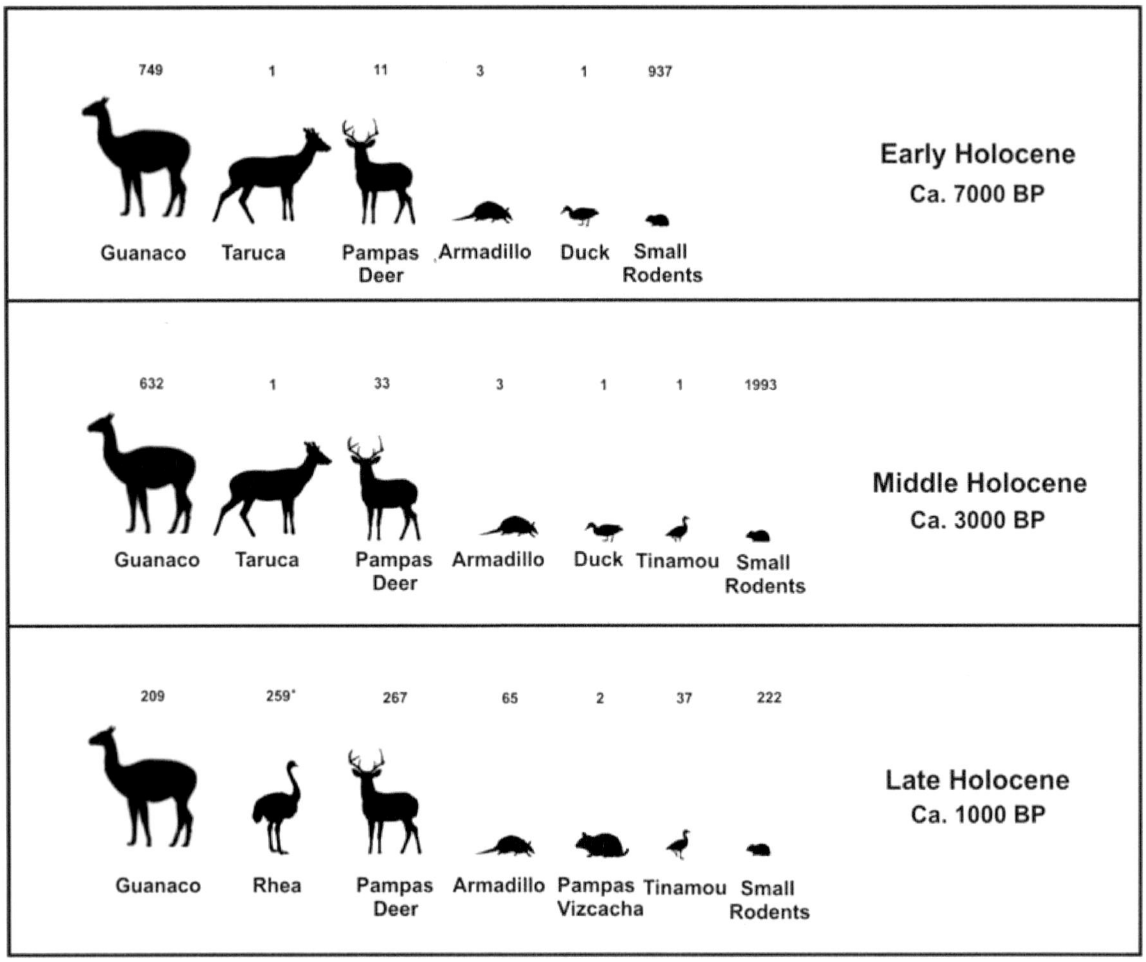

FIGURE 2. NUMBER OF IDENTIFIED SPECIMENS PER TAXON (NISP) FOR THE UPPER MOUNTAIN GRASSLAND RANGE ASSEMBLAGES (CÓRDOBA PROVINCE, ARGENTINA).

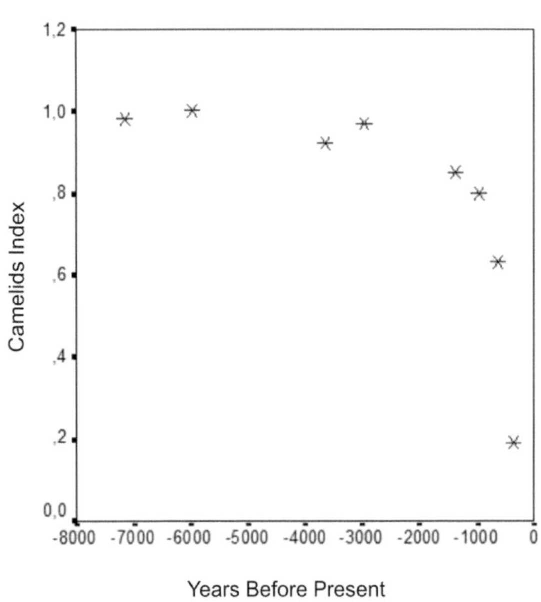

FIGURE 3. TEMPORAL TRENDS IN CAMELIDS ABUNDANCE FOR THE UPPER MOUNTAIN GRASSLAND RANGE ASSEMBLAGES (CÓRDOBA PROVINCE, ARGENTINA).

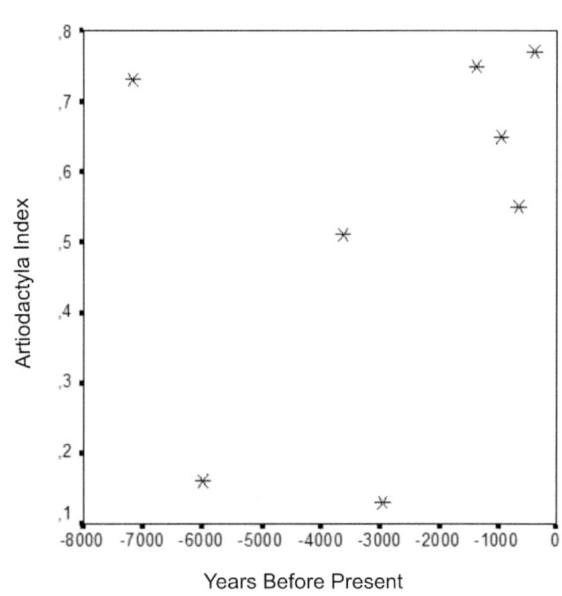

FIGURE 4. TEMPORAL TRENDS IN ARTIODACTYLA ABUNDANCE FOR THE UPPER MOUNTAIN GRASSLAND RANGE ASSEMBLAGES (CÓRDOBA PROVINCE, ARGENTINA).

Site	Date (BP)	Camelids Index	Artiodactyla Index
AEG1-C1	7160	0,98	0,73
QR1-C1	5980	1	0,16
AEG1-C2	3637	0,92	0,51
QR1-C2	2950	0,97	0,13
RY-11	1361	0,85	0,75
LA-1	949	0,8	0,65
RY-14	640	0,63	0,55
PE-1	364	0,19	0,77

Table 6. Results of the "camelids index" and the "artiodactyla index" for the upper mountain grassland range assemblages (Córdoba Province, Argentina).

Index	Chi-Cuadrado	Valor *p*
Camelids : Deers	1454,44	0,000
Artiodactyla : Rodents	2130,52	0,000

Table 7. Chi-Square Test linear trends for the upper mountain grassland range assemblages (Córdoba Province, Argentina).

The optimal diet breadth never predicts the anthropic predation of small-vertebrates. Nevertheless, caviomorph-rodents (*Lagostomus maximus, Ctenomys* sp., Caviinae, etc.), cricetid-rodents (*Holochilus* sp.) and birds (Tinamidae, *Anas* sp., etc.) are found in the assemblages with unequivocal traces of human interaction such as cut-marks and burning damage. Two hypothesis would explain their archaeological presence:

Hypothesis 1: Participation of women and children in the economy

Ethnoarchaeological and ethnographic studies in small-scale societies pointed out that subsistence activities are divided according to sex and age differences. Women and children do not participate directly in adult men large-game hunting, although they usually colaborate with the processing and transportation of the prey body parts to the base camp. Their economical participation focuses on the daily provision of food foraging low-return and easy to capture resources, such as small-animals, fruits and tubers (Zeanah 2004; Byers & Ugan 2005 Prates 2008; Bird & Bliege Bird 2000; Wadley 2009). Thus, the identification of small-vertebrates in the sites, specially in Middle Holocene deposit, might be related to the archaeological consequence of these sexual and age-linked foraging practices.

Hypothesis 2: Problems in the population-resource relationship

An alternative explanation argues that the small animals were exploited as a result of anthropogenic reduction in high-ranked prey density. Thus, a prey simulation model was used to examine the long-term stability of a human subsistence system based on *L. guanicoe*, the high-ranked but susceptible to over-explotation prey type in Sierras of Cordoba (Medina and Rivero 2007). The simulation employed biological and ecological variables of *L. guanicoe* (e.g. weight, density, population dynamics) and humans populations (e.g. calory requirements, demographic growth) to predicts how these model parameters behave when one of them was altered (e.g. population density). Medina and Rivero´s (2007) model predicts that one-prey subsistence system focused on *L. guanicoe* should not be maintainable as far from the Middle Holocene ca. 6000 BP. At this threshold, habitat fragmentation and human predation produces a dramatic decrease in their local abundance. Human forager should respond by reducing prey selectivity and adding small-vertebrates that normally were marginally outside of the optimal diet (Rivero 2009; Medina & Rivero 2007).

According to this hypothesis, the declining trend in *L. guanicoe* prey density and the increase of the diet breadth constitute an obvious case of niche construction because humans alter the selective pressure through their own cultural activity and drive further adaptive adjustments. Thus, the predictions of Medina and Rivero (2007) prey simulation models resemble the "*Route* 1" path of Odling-Smee et al (2003: 338), in which the socioecological environment modified by cultural niche construction directs the evolutionary dynamics of the human societies.

CONCLUSIONS

The archaeofaunal record from the Sierras of Cordoba upper mountain grassland range reflect the temporal decrease in *Lama* sp. frequency through the Holocene. This taxonomical trend is associated here to a decrease in their encounter rate as result of human over-exploitation under specific ecological conditions. Conversely, Odocoleinae increase their abundance during this period and their importance in the subsistence is greater than *Lama* sp. towards the Late Holocene. Small-rodents were present at low-NISP values since Early Holocene, but their frequency increases significantly from Middle to Late Holocene faunal assemblages. This pattern cannot be explained only with the Hypothesis 1 and is interpreted as a human behavioral adjustment to the *L. guanicoe* prey depression in agreement with Hypothesis 2.

However, only a preliminary evaluation of the model predictions was carried out in this chapter. The assessment of the model performance requires to eliminate the potencially sampling bias and the chronological gaps –including the absence of pleistocene faunal assemblages- through regional-scale archaeological studies. Furthermore, the analysis of the predator-prey dynamics and the evolutionary history of the human societies should be integrated with non-faunal evidence to expand its explanatory power. Thus, the changes in projectile point design, the development of new technologies –e.g. grinding tools, pottery, etc.–, the frequency of archaeobotanical remains and the variability in home range size mobility should also be considered, with special attention to their effects on foraging efficiency, demography, mobility, territorial defence, social differentiation, palaeopathology, etc.

The chapter also points out how the neo-darwinian evolutionary logic allows the operationalization of highly-predictive models about human colonization process and its effect on the ungulate prey resource they exploit. This is the opposite to the non-evolutionary approaches which claim for an unwarranted equilibrium between humans and the environment (Redman 1999; Smith & Wishnie 2000; Winterhalder & Lu 1997). The *Homo sapiens* spread through the evolutionary scenarios of Argentina should promote co-evolutionary dynamics similar to the ones proposed here (Kelly 1999; Winterhalder 1981b). Nevertheless, different results from local extinction –e.g. domestication, dispersion of species, high-efficiency subsistence systems based on Camelidae, etc. (López 2009; Stahl 2009; Yacobaccio 2001)- should be considered due to the variability in the specific socio-ecological settings.

The archaeology of the Sierras of Cordoba should advance in the obtention of new faunal assemblages to test the model predictions with a high empirical robustness. However, the available archaeological data supports a broad-spectrum diet which may not be an exclusively Late Holocene phenomenon and its origins may be found in earlier processes beginning during the Middle Holocene. Thus, the incorporation of small-scale agriculture around *ca.* 1200-1000 BP cannot be satisfactorily explained only by external factors –e.g. climate changes, the diffusion and/or arrival of new biological populations, etc.-. A niche construction conception of evolution as a long-term behavioral cumulative process should also be taken into account. Indeed, research should consider that past human cultural activities played a dynamic role in this process through niche construction, co-directing subsequent cultural changes that are seen in the archaeological record.

Acknowledgements

We thank Consejo Nacional de Investigaciones Científicas y Técnicas (CONICET) for the financial support (PIP 11220080102678) and to Eduardo Berberián for his kind professional advice and help. Our acknowledgement also extends to Sebastian Pastor, Hernán Muscio, Pablo Teta and Laura López who provided bibliography and replied to our numerous requests to improve the original manuscript.

References

Adams, J. and H. Faure (Eds.) 1997. *QEN members. Review and Atlas of Palaeovegetation: Preliminary land ecosystem maps of the world since the Last Glacial Maximum.* Oak Ridge National Laboratory, USA. http://www.esd.ornl.gov/projects/qen/adams1.html

Berberián, E.; S. Pastor; D. Rivero; M. Medina; A. Recalde; L. López and F. Roldán 2008. Últimos avances de la investigación arqueológica en las Sierras de Córdoba. *Comechingonia* 11: 135-164.

Bettinger, R. 1991. *Hunter-gatherers: Archaeological and Evolutionary Theory.* Plenum Press, New York.

Bird, D. and R. Bliege Bird 2000. The Ethnoarchaeology of Juvenile Foragers: Shellfishing Strategies among Meriam Children. *Journal of Anthropological Archaeology* 19:461-476 .

Bird, D. and J. O´Connell 2006. Behavioral Ecology and Archaeology. *Journal of Archaeological Research* 14: 143-188

Boone, J. and E. Smith 1998. Is it the evolution yet? A critique of Evolutionary Archaeology. *Current Anthropology* 39: 141-173.

Borrero, L. 1994-1995. Arqueología de la Patagonia. *Palimpsesto* 4: 9-69.

Broughton, J. 1994. Late Holocene resource intensification in the Sacramento Valley, California: the vertebrate evidence. *Journal of Archaeological Science* 21: 501-514.

Broughton, J. 1999. *Resource Depression and Intensification during the Late Holocene, San Francisco Bay. Evidence from the Emeryville Shellmound Vertebrate Fauna.* University of California Press, California.

Broughton, J.; M. Cannon and E. Bartelink 2010. Evolutionary Ecology, Resource Depression, and Niche Construction Theory: Application to Central California hunter-gatheres and Mimbres-Mogollon agriculturalist. *Journal of Archaeological Method and Theory*, in press.

Broughton, J. and D. Grayson 1993. Diet breath, adaptative change and the White Mountains Faunas. *Journal of Archaeological Science* 20: 331-336.

Bucher, E. and J. Abalos 1979. Fauna. In *Geografía Física de la Provincia de Córdoba.* Edited by J. Vázquez, R. Miatello and M. Roqué pp. 45-138. Editorial Boldt, Buenos Aires.

Byers, D. and A. Ugan 2005. Should we expect large game specialization in the Late Pleistocene? An optimal foraging perspective on Early Paleoindian prey choice. *Journal of Archaeological Science* 32:1624-1640.

Cabido, M.; G. Funes; E. Pucheta; F. Vendramini and S. Díaz 1998. A chorological análisis of the mountains from Central Argentina. Is all what we call Sierra Chaco really Chaco? Contribution to the study of the flora and vegetation of the Chaco. *Candollea* 53: 321-331.

Carignano, C. 1999. Late Pleistocene to recent climate change in Córdoba Province, Argentina: geomorphological evidence. *Quaternary International* 57/58: 117–134

Cei, J. 1972. Segregación corológica y procesos de especiación por aislamiento en anfibios de Pampa de Achala, Córdoba. *Acta Zoológica Lilloana* XXIX: 233-245.

Demaio, P.; O. Karlin and M. Medina 2002. *Arboles Nativos del Centro de la Argentina.* L.O.L.A. (Literature of Latin American), Buenos Aires.

González, A.R. 1960. La estratigrafía de la gruta de Intihuasi (Prov. de San Luis, R.A.) y sus relaciones con otros sitios precerámicos de Sudamérica. *Revista del Instituto de Antropología* I: 5-296.

Grayson, D. 2001. The archaeological record of human impact on animal populations. *Journal of World Prehistory* 15 (1): 1-68.

Grayson, D.; F. Delpech; J.P. Rigaud and J. Simek 2001. Explaining the development of dietary dominance by a single ungulate taxon al Grotte XVI, Dordogne, France. *Journal of Archaeological Science* 28: 115-125.

Hill, K. and K. Hawkes 1983. Neotropical hunting among the Aché of Eastern Paraguay. In *Adaptative Responses of Native Amazonian.* Edited by R. Hames and W. Vickers, pp. 139-188. Academic Press, New York.

Kaplan, H. and K. Hill 1992. The Evolutionary Ecology of food acquisition. In *Evolutionary Ecology and Human Behavior* . Edited by E. Smith and B. Winterhalder, pp. 167-201. Aldine de Gruyter, New York.

Kelly, R. 1995. *The Foraging Spectrum. Diversity in Hunter-Gatherer Lifeways*. Smithsonian Institution Press, Washington.

Kelly, R. 1999. Hunter-gatherer foraging and colonization of the Western Hemisphere. *Anthropologie* 37:143-153

Laland, K.; J. Odling-Smee and M. Feldman 2001. Cultural niche construction and human evolution. *Journal of Evolutionary Biology* 14:22-33.

López, G. 2009. Arqueofaunas, osteometría y evidencia artefactual en Pastos Grandes, Puna de Salta: secuencia de cambio a lo largo del Holoceno Temprano, Medio y Tardío en el sitio Alero Cuevas. *Intersecciones en Antropología* 10: 105-119.

Luti, R.; M. Bertran de Solís; F. Galera; N. Muller de Ferreira; M. Berzal; M. Nores; M. Herrera and J. Barrera 1979. Vegetación. In *Geografía Física de la Provincia de Córdoba*. Edited by J. Vázquez, R. Miatello and M. Roqué pp. 297-368. Editorial Boldt, Buenos Aires.

Lyman, R. 2008. *Quantitative Paleozoology*. Cambridge University Press, Cambridge.

Lyman, R. and K Cannon 2004. Applied zooarchaeology, because it matters. In *Zooarchaeology and Conservation Biology*. Edited by R. Lyman and K. Cannon, pp. 1-24, The University of Utah Press, Salt Lake City.

Medina, M. 2008. Diversificación Económica y Uso del Espacio en el Tardío Prehispánico del Norte del Valle de Punilla, Pampa de Olaen y Llanura Noroccidental (Córdoba, Argentina). Unpublished Ph.D. Dissertation. Facultad de Filosofía y Letras, Universidad de Buenos Aires, Buenos Aires. Ms.

Medina, M. 2009. Tendencias en el consumo prehispánico tardío de recursos faunísticos: zooarqueología de C.Pun.39 y Puesto La Esquina 1 (Córdoba, Argentina). *Archaeofauna, International Journal of Archaeozoology* 18, 119-136.

Medina, M., Rivero, D. 2007. Zooarqueología, *Lama guanicoe* y dinámica evolutiva del Chaco Serrano. *Mundo de Antes* 5, 211-234.

Medina, M.; Rivero, D. and P. Teta 2009. Consumo antrópico de pequeños mamíferos en el Holoceno de Argentina central: perspectivas desde el abrigo rocoso Quebrada del Real 1 (Pampa de Achala, Córdoba). *Latin American Antiquity*, in press.

Medina, M.; L. Turnes and E. Apolinaire 2010. La explotación de huevos de Rheidae en el sitio prehispánico tardío Arroyo Talainín 2 (Sierras de Córdoba, Argentina). Paper presented at the XVII Congreso Nacional de Arqueología Argentina, Mendoza.

Menghin, O. and A. González 1954. Excavaciones arqueológicas en el yacimiento de Ongamira, Córdoba (Rep. Arg.). Nota preliminar. *Notas del Museo de La Plata*, tomo XVII, Antropología n° 67: 213-274.

Muscio, H. 1999. Colonización humana del NOA y variación en el consumo de los recursos: La ecología de los cazadores recolectores de la Puna durante la Transición Pleistoceno-Holoceno. *Revista Naya http//www.naya.org.ar/articulos/index.htm*

Muscio, H. 2004. Dinámica Poblacional y Evolución Durante el Período Agroalfarero Temprano en el Valle de San Antonio de los Cobres, Puna de Salta, Argentina. Unpublished Ph.D. Dissertation. Facultad de Filosofía y Letras, Universidad de Buenos Aires, Buenos Aires. Ms

Nagaoka, L. 2002. The effects of resource depression on foraging efficiency, diet breadth, and patch use in Southern New Zealand. *Journal of Archaeological Science* 21: 419-442.

O'Connell, J. 1995. Ethnoarchaeology needs a general theory of behavior. *Journal of Archaeological Research* 3(3): 205-255.

Odling-Smee, J. 2007. Niche inheritance: A possible basis for classifying multiple inheritance system in human evolution. *Biological Theory* 2(3):276-289.

Odling-Smee, J.; k. Laland and M. Feldman 2003. *Niche Construction. The Neglected Process in Evolution*. Princeton University Press, Princeton.

Pastor, S. 2005. El sitio Río Yuspe 14 (Pampa de Achala, Córdoba). Perspectivas sobre el uso prehispánico tardío de los ambientes serranos de altura. *Mundo de Antes* 4: 87-104.

Pastor, S. 2007a. Arqueología del valle de Salsacate y Pampas de Altura Adyacentes (Sierras Centrales de Argentina). Una Aproximación a los Procesos Sociales del Período Prehispánico Tardío (900-1573 d.C.). Unpublished Ph.D. Dissertation. Facultad de Ciencias Naturales y Museo, Universidad Nacional de La Plata, La Plata. Ms

Pastor, S. 2007b. "Juntas y cazaderos". Las actividades grupales y la reproducción de las sociedades prehispánicas de las Sierras Centrales de Argentina. In *Procesos Sociales Prehispánicos en el Sur Andino: la Vivienda, la Comunidad y el Territorio*. Edited by A. Nielsen, V. Seldes, M. Vázquez and P. Mercolli, pp. 361-376. Editorial Brujas, Córdoba.

Piovano, E.; D. Ariztegui; F. Córdoba; M. Cioccale and F. Sylvestre 2009. Hydrological variability in South America below the tropic of Capricorn (Pampas and eastern Patagonia, Argentina) during the last 13.0 ka. In: *Past Climate Variability in South America and Surrounding Region*. Edited by F. Vimeux, F. Sylvestre and M. Khodri. *Developments in Paleoenvironmental Research* 14: 323-351. Springer, Dordrecht, The Netherlands.

Polop, J. 1989. Distribution and ecological observations of wild rodents in Pampa de Achala, Córdoba. *Studies in Neotropical Fauna and Environments*, 24: 53-59.

Prates, L. 2008. *Los Indígenas del Rio Negro. Un Enfoque Arqueológico*. Sociedad Argentina de Antropología, Buenos Aires.

Redman, C. 1999. *Human Impact on Ancient Environments*. University of Arizona Press, Tucson

Recalde, A., 2009. Las Representaciones Rupestres en el Sector Occidental de las Sierras Centrales y su Relación con las Estrategias de Explotación de Recursos de las Comunidades Prehispánicas Productoras de Alimentos (Pcia. de Córdoba). Unpublished Ph.D. Dissertation. Facultad de Filosofía y Humanidades, Universidad Nacional de Córdoba, Córdoba. Ms.

Ringuelet, R. 1961. Rasgos fundamentales de la zoogeografía de la Argentina. *Physis* 63: 151-170.

Rivero, D. 2009. *Ecología de Cazadores-Recolectores del Sector Central de las Sierras de Córdoba (Rep. Argentina)*. BAR Internacional Series 2007, Oxford.

Rivero, D. and F. Roldán 2005. Inicial peopling of the Córdoba Mountains, Argentina: first evidence from El Alto 3. *Current Research in The Pleistocene* 22: 33-35.

Rivero, D. and E. Berberián 2008. El poblamiento de la región central del territorio argentino durante la Transición Pleistoceno-Holoceno (12000 - 9000 A.P.). *Revista Española de Antropología Americana* 38(2):17-37.

Rivero, D; S. Pastor and M. Medina 2010. Intensificación en las Sierras de Córdoba. El Abrigo Rocoso Quebrada del Real 1 (ca. 6000-500 AP, Córdoba, Argentina). *Anales de Arqueología y Etnología de la Universidad Nacional de Cuyo* 63-64:227-246.

Rivero, D.; M. Medina; A. Recalde and S. Pastor 2010. Variabilidad en la explotación de recursos faunísticos durante el Holoceno en las Sierras de Córdoba (Argentina): una aproximación zooarqueológica. In *Zooarqueología a Principios del Siglo XXI: Aportes Teóricos, Metodológicos y Casos de Estudio*. Edited by M. Gutierrez, M. De Nigris, P. Fernández, M. Giardina, A. Gil, A. Izeta, G. Neme and H. Yacobaccio, pp. 321-331. Ediciones El Espinillo, Buenos Aires.

Sanabria, J. and G. Arguello 2003. Aspectos geomorfológicos y estratigráficos en la génesis y evolución de la Depresión Periférica, Córdoba (Argentina). *Actas II Congreso Argentino de Cuaternario y Geomorfología (Tucumán)*: 177-184.

Smith, E. 1983. Anthropological application of Optimal Foraging Theory: a critical Review. *Current Anthropology* 24(5): 625-650.

Smith, E. and M. Wishnie 2000. Conservation and subsistence in small-scale societies. *Annual Review of Anthropology* 29: 493-524.

Smith, B. 2007. Niche construction and behavioral context of plant and animal domestication. *Evolutionary Anthropology* 16:188-199

Stahl, P. 2009. Adventive vertebrates and Historical Ecology in the Pre-Columbian Neotropics. *Diversity* 1: 151-165;

Stephen, D. and J. Krebs 1986. *Foraging Theory*. Princeton University Press, Princeton.

Stiner, M. 1994. *Honor Among Thieves. A Zooarchaeological Study of Neardental Ecology*. Princeton University Press, Princeton.

Teta, P.; M. Medina; S. Pastor; D. Rivero and H. Paradela 2005. Holochilus brasiliensis (Rodentia, Cricetidae) en conjuntos arqueofaunísticos del Holoceno Tardío de la Provincia de Córdoba (Argentina). *Mastozoología Neotropical* 12(2): 271-275.

Wadley, L. 2009. Were snares and traps used in the Middle Stone Age and does it matter? A review and a case study from Sibudu, South Africa. *Journal of Human Evolution* 58 (2010) 179–192.

Winterhalder, B. 1981a. Optimal foraging strategies and hunter-gatherers research in Anthropology: theory and models. In *Hunter-Gatherer Foraging Strategies: Ethnographic and Archaeological Analyses*. Edited by B. Winterhalder and E. Smith, pp. 13-35. University of Chicago Press, Chicago.

Winterhalder, B. 1981b. Foraging Strategies in the Boreal Forest: an analisis of Cree hunting and gathrering. In *Hunter-Gatherer Foraging Strategies: Ethnographic and Archaeological Analyses*. Edited by B. Winterhalder and E. Smith, pp. 66-98. University of Chicago Press, Chicago.

Winterhalder, B. and F. Lu 1997. A Forager-Resource Population Ecology Model and Implication for Indigenous Conservation. *Conservation Biology* 11 (6): 1354-1364.

Winterhalder, B. and E. Smith. 1992. Evolutionary Ecology and the Social Science. In *Evolutionary Ecology and Human Behavior*. Edited by E. Smith and B. Winterhalder, pp. 3-23. Aldine de Gruyter, New York.

Winterhalder, B. and E. Smith. 2000. Analysing Adaptative Strategies: Human Behavioral Ecology at Twenty-Five. *Evolutionary Anthropology* 9(2): 51-72.

Yacobaccio, H. 2001. La domesticación de camélidos en el Noroeste Argentino. In *Historia Argentina Prehispánica.* Edited by E. Berberián and A. Nielsen, pp. 7-40. Editorial Brujas. Córdoba.

Yacobaccio, H.; C. Madero and M. Malmierca 1998. *Etnoarqueología de Pastores Surandinos*. Grupo de Zooarqueología de Camélidos, Bs. As.

Zeanah, D. 2004. Sexual divisón of labor and Central Place Foraging. A model for the Carson Desert of Western Nevada. *Journal of Anthropological Archaeology* 23:1-32.

The Development of a Legacy: Evolution, Biogeography and Archaeological Landscapes

Juan Bautista Belardi[1], Ramiro Barberena[2], Rafael Goñi [3] and Anahi Re[4]

1 Universidad Nacional de la Patagonia Austral. Unidad Académica Río Gallegos. CONICET. Campus Universitario, Avda. Gregores y Piloto Lero Rivera. (9400) Río Gallegos, Santa Cruz, Argentina. silespi@infovia.com.ar
2 CONICET-IMHICIHU-DIPA. Saavedra 15, 5to (1083) Ciudad Autónoma de Buenos Aires, Argentina. Buenos Aires. ramidus28@gmail.com 3 Instituto Nacional de Antropología y Pensamiento Latinoamericano. Universidad de Buenos Aires y Universidad Nacional del Centro de la Provincia de Buenos Aires.
3 de Febrero 1370 (1426) Ciudad Autónoma de Buenos Aires, Argentina. gonirafael@gmail.com
4 Instituto Nacional de Antropología y Pensamiento Latinoamericano. CONICET. 3 de Febrero 1370 (1426) Ciudad Autónoma de Buenos Aires, Argentina. anahire1@gmail.com

Abstract

The Darwinian theory of evolution has a key incidence in different levels of archaeological practice that constitute a research program. This goes from a basically theoretical instance, associated with several frameworks in fashion during the last decades to a mid-range level, which is composed of sets of basically complementary models and frames of reference: optimal foraging theory, cultural transmission, costly signalling, niche construction, all of which imply an archaeological assessment in a regional scale. The main challenge that we face is to make these models -derived from different theoretical strands- operative at the level of the archaeological record. The aid of wide spatial and temporal analytical scales is as simple as fundamental by allowing the identification of material variation. The approaches of biogeography and archaeological landscapes provide a crucial aid in this regard. Biogeography may be regarded as a mid-range instance necessary to operationalize the archaeological analysis of evolutionary processes. In this context, human populations are seen as components of ecosystems and space is the dimension where the intersection between models of cultural and genetic transmission is evaluated. Our goal is to introduce a brief archaeological discussion in a biogeographical frame exploring the relationship between some models derived from evolutionary theory. We define cultural transmission as a part of niche construction theory, allowing relating spatial and temporal trajectories of human behaviour. Firstly we introduce the theoretical concepts that guide our work, focusing on the relationship between evolution and biogeography, archaeological landscapes and cultural transmission. Secondly, these relations are illustrated through the development of two case studies from Patagonia, which focus on the distribution of rock-art motifs and may be evaluated and discussed in the light of these concepts.

Resumen

La teoría darwiniana de la evolución tiene una incidencia importante en los diferentes niveles de la práctica arqueológica, que constituyen un programa de investigación. Esto va desde una instancia básicamente teórica, relacionada con distintos modelos desarrollados en las últimas décadas a un nivel de rango medio, compuesto por conjuntos de modelos complementarios y marcos de referencia: la teoría de forrajeo óptimo, la transmisión cultural, la señalización costosa, la construcción nicho, todo lo cual implica dar cuenta del registro arqueológico en una escala regional. El principal desafío al que nos enfrentamos es hacer estos modelos, -derivados de diferentes líneas teóricas- operativos al nivel del registro arqueológico. El empleo de escalas espaciales y temporales amplias, es tan simple como fundamental ya que permite identificar la variación material. Los enfoques de la biogeografía y los paisajes arqueológicos ofrecen una herramienta esencial a este respecto. La biogeografía puede ser considerada como una instancia de rango medio necesario para operacionalizar el análisis arqueológico de los procesos evolutivos. En este contexto, las poblaciones humanas son vistas como componentes de los ecosistemas y el espacio es la dimensión en la que se evalúa la intersección entre los modelos de transmisión cultural y genética. Nuestro objetivo es presentar una breve discusión arqueológica en un marco biogeográfico y explorar la relación entre algunos modelos derivados de la teoría de la evolución. Definimos la transmisión cultural como parte de la teoría de construcción de nicho, lo cual permite relacionar las trayectorias espaciales y temporales observadas en la conducta humana. En primer lugar se introducen los conceptos teóricos que guían nuestro trabajo, centrándonos en la relación entre la evolución y biogeografía, paisajes arqueológicos y transmisión cultural. En segundo lugar, estas relaciones se ilustran a través de dos caso de estudios de la Patagonia, centrados en la distribución de los motivos rupestres; los cuales son evaluados y discutidos a la luz de estos conceptos.

Introduction

In the last chapters of Charles Darwin's work (1859), "On the Origins of Species by means of natural selection or the preservation of favoured races in the struggle for life", there are concepts of main importance for the work that we pursue here. While chapters IX and X focus on the geological record and the organisms, chapters XI and XII focus on their geographical distribution. Thus, Darwin's main legacy was clearly established: the relationship between time, space and change based on natural selection. One hundred and fifty years later it is possible to see a close relationship between concepts derived from evolutionary theory, the development of an archaeology that is defined by the study of wide spaces (Foley 1981; Borrero 1993a) and the consideration of the analytical potential offered by the study of temporal trajectories as revealed in the archaeological record.

The Darwinian theory of evolution has a key incidence in different levels of archaeological practice that constitute a research program in Imre Lakato's terms (see Scheinsohn 2008). This goes from a basically theoretical instance, associated with several frameworks in fashion during the last decades (Rindos 1980; O'Connell 1995; Bettinger 2008), to a mid-range level or "protecting belt" of the theory's core, which is composed of sets of basically complementary models and frames of reference: optimal foraging theory, cultural transmission, costly signalling, niche construction, among others, all of which imply an archaeological assesssment in a regional scale. Even the analyses centered in archaeological sites refer to space as a bigger whole, such as in the case of the so-called 'monitoring perspective' (Thomas 1983). The analysis of movement costs, selection of preys, raw materials and their processing and transportation costs, as well as the presence of signal emitters and receptors, are all integrated around the notion of behaviour as spatially continuous. The main challenge that we face, as it is widely known, is to make these models -derived from different theoretical strands- operative at the level of the archeological record (v.g. Lyman 2003; Eerkens and Lipo 2007). The aid of wide spatial and temporal analytical scales is as simple as fundamental by allowing the identification of material variation, which is the raw matter these models work with.

We consider that the approaches in the fields of biogeography and archaeological landscapes provide a crucial aid in this regard (Borrero 2005). Biogeography, which acquires an explicit theoretical character in the pioneer works of Darwin and Alfred Wallace (Briggs and Humphries 2004), provides concepts that allow gathering the spatial dimension with specific trajectories of cutural and biological change, constituting the basis for important archaeological developments (Borrero 1989-1990; Veth 1993). Biogeography may be regarded as a mid-range instance necessary to operationalize the archaeological analysis of evolutionary processes. In this context, human populations are seen as components of ecosystems (Pianka 1982) and space is the dimension where the intersection between models of cultural and genetic transmission is evaluated.

Our goal in this paper is to introduce a brief archeological discussion in a biogeographical frame exploring the relationship between some models derived from evolutionary theory. We are particularly seeking to contemplate cultural transmission as a part of niche construction theory (Aunger 2009; see general frame in Odling-Smee et al. 2003), allowing relating spatial and temporal trajectories of human behaviour. Firstly we introduce the theoretical concepts that guide our work, focusing on the relationship between evolution and biogeography, archaeological landscapes and cultural transmission. Secondly, these relations are illustrated through the development of two case studies from Patagonia (Figure 1), which focus on the distribution of rock-art motifs and may be evaluated and discussed in the light of these concepts.

Evolution and biogeography

From its beginnings as a discipline, biogeography has provided the necessary connection between evolution and space. Biogeography may be briefly characterized as an interdisciplinary field of research that allows connecting the distribution of a set of organisms -past or present- with a landscape that has a given morphological and ecological structure that changes throughout time. All speciation models require space as a main element in their explanatory arguments, whether they emphasize sympatric or alopatric mechanisms and, in the latter case, vicariant or dispersion processes (Giller et al. 2004). Biogeography and the niche construction frame are both derived from evolutionary theory and share a number of principles and goals. In both cases, the structure and history of a landscape are key issues.

Niche construction theory sets the organisms as shaping forces of the habitat they occupy, thus contributing to the formation of future selective environments for their own species as much as for others (Odling-Smee et al. 2003).

Since populations are conceived as components and shaping forces of the ecosystems they inhabit, and behaviour is conceived as spatially continuous (Foley 1981), biogeographical models connect a given landscape with the demographical trajectories of a population (such as the model proposed by Veth [1993] for Central Australia, which suggests the existence of corridors, barriers and refugees).

By doing this, they provide a frame to evaluate the transmission of cultural and biological information with a key place for the analysis of temporal trajectories of change. The divergent evolution model proposed by Borrero (1989-1990) for southern Patagonia provides another excellent example of this perspective. In synthesis, biogeographical models offer a spatial frame to interpret long-term biological and cultural change.

Evolution, Archaeological Landscapes and Cultural Transmission

An archeological landscape is understood as the attritional product of human occupations and formation processes in a given space through time (see Yacobaccio 1988). There are two dimensions that become of great interest here: the first derives from niche construction theory and suggests that the landscape is the product of the interaction between organisms and their environment that is inherited with a set of signals and modifications taking place in the long run (Odling-Smee et al. 2003; see Muscio 2009). In this context, a distinction formulated by R. Aunger becomes relevant, who differentiates between two ways of social learning that carry different types of environmental modification.

The first is related to the concept of communication defined as "an instance of niche construction using signs, signals, or artifacts that is targeted at changing the behaviour of conspecifics", which implies intentional modifications of the environment (Aunger 2009:

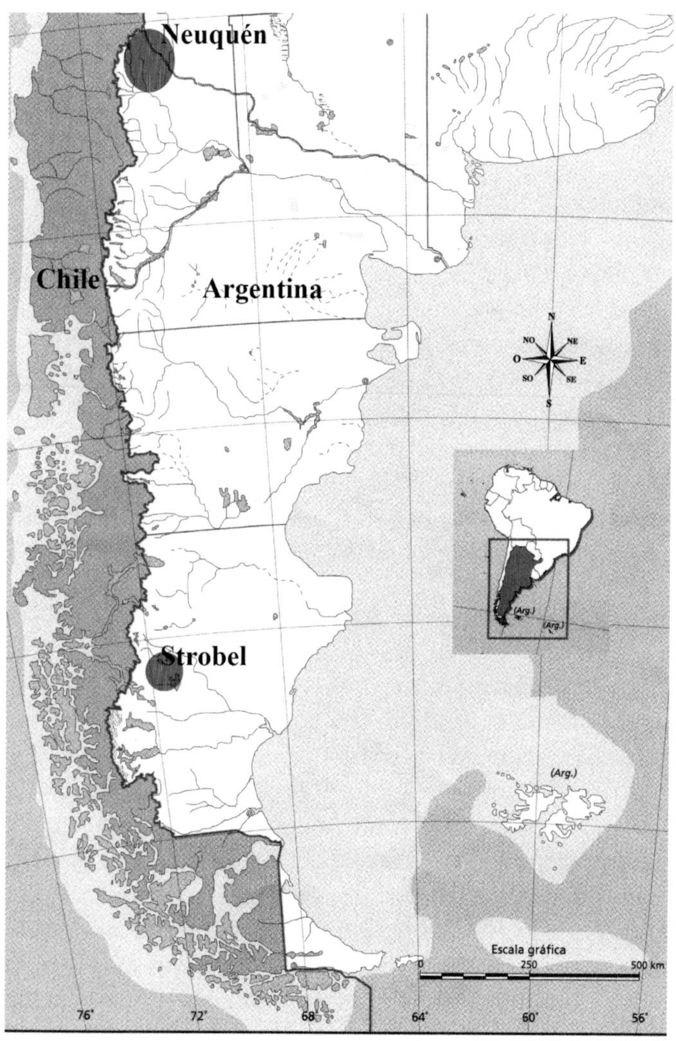

FIGURE 1. ARCHAEOLOGICAL CASE-STUDIES FROM PATAGONIA

37). The simplest form of communication refers to the production of signals as verbalizations or body movements that do not have an archaeological correlate. Logically, there are also forms of communication that imply the formation of a material record and which generate the possibility of storing information within a regional landscape that may be 'read' afterwards. Following this idea, rock-art representations clearly constitute an intentional modification of the landscape and may be seen as an instance of niche construction through the transmission of information. At the same time, rock-art representations, in contrast with body or verbal signals, have the capacity of being received in a delayed form.

On the other hand, Aunger (2009) identifies another form of social learning that involves the unintentional transmission of information, including the observation of the behaviour of conspecifics, as well as the incidental changes generated in the landscape through the behaviour of an individual or population. Even though an intentional landscape modification is important because it may highlight the 'strategic' nature of certain instances of niche construction (in Aunger's sense), it does not cover all the importance of this phenomenon. As Martínez and Mackie suggest (2003-2004) when analizing the lithification of Quequén River's basin, the consequences of certain activities in later human behaviour may be independent of their intentional vs. incidental character.

Wobst (1977) and Wiessner's (1983) suggestions may be rethought within this perspective. It is useful to consider certain characteristics proposed by Wobst for communication through the production of artifacts: the message may be transmitted in the absence of receptors; in this context the message changes more slowly and implies higher production costs, even though it requires less maintenance costs and may potentially reach a larger amount of receptors. Then, intentional modification of the landscape may have a very important role and, once a given space is effectively integrated within nets of mobility and circulation of material objects and information, they may be crucial to evaluate the interaction between populations (Belardi and Goñi 2006; Taçon 2008).

Another aspect that we consider of interest is related to cultural transmission models, which have an important place in the landscape learning process (Rockman and Steele 2003), including the different resources necessary for the subsistence and reproduction of a society. Relating to the mid-range level mentioned above, this shows a connection with optimal foraging models of behaviourial ecology and with dual-inheritance theory (Richerson and Boyd 2005; Bettinger 2008; Shennan 2009). As Bettinger (2008: 1-2) suggests, the combination of these different perspectives allows to place the analysis of cultural change in a frame that evaluates the information available in a system, limiting the extent of the assumption of complete information availability in the level of the individual (a basic point in optimal foraging models). This knowledge of the landscape has an influence on later occupations and creates restrictions to adaptability, being its evaluation possible by means of the study of archaeological landscapes.

Evolution and Temporal Trajectories

The diachronic assessment of archaeological landscapes is a key issue. Temporal trajactories in the expression of several proxies provide the basis to discuss cultural change that may or may not be adaptive for human populations (Borrero 1993a). A first step is to discriminate the existence of "differential positioning of systems" (Binford 1983) as regards intrinsic change within them. In this way, the best strategy is the integration of independent evidences that may be mutually contrasted. Regional space is the scale where these subjects should be evaluated by the analysis of variables like intensity and redundance of use, which are meassurable distributionally on the basis of deposition rates (Borrero 1993b; Belardi 2005; Barberena 2008; Hiscock 2008) that provide an estimate of the hierarchy in the use of space. The continuities-discontinuities in landscape use are directly related to these concepts because of the need of new instances of environmental learning.

To evaluate the precedent situations, several scenarios where rock-art representations are included at the artifactual level may be modelled. We may think of a situation in which there is continuity or change in the human use of a space by a given population or, on the other hand, where there is a discontinuity at a population level (Figure 2). Thus, case 1 represents a space occupied by the same population and under the same form of use through time. This might be seen as a form of persistence or stasis. In case 2, change takes place at the level of human populations, even though the form of use of space is similar, which may happen under conditions of environmental stability. Change in the use of space but not in the population level takes place in case 3. In this case, variation should be observed in the archeological record and is clearly a configurational change in a given human system. Lastly, in case 4, change takes place in both the population level and in the the form of use of the space. Here, environmental conditions and the economic and political organization of the populations involved may have varied. This simple grid provides a frame for the evaluation of temporal discontinuities with a probable demographic base (Barrientos 2009; Martínez 2008-2009; see also Zeanah and Simms 1999). At the same time, it allows to advance in the analysis of artifactual changes. In the case of rock-art evidences, changes in the density and diversity of motifs and designs, manufacture techniques and bedrock employed can be monitored.

Patterns in hunter gatherer space use		
Population	Same	Different
Same	Case1	Case 3
Different	Case 2	Case 4

FIGURE 2. DIACHRONIC RELATIONSHIPS BETWEEN POPULATIONS AND ARCHAEOLOGICAL LANDSCAPES FOR A HYPOTHETICAL SPACE.

These aspects may be evaluated on the basis ofhuman remains, artifactual distributions and radiocarbon datasets, and necessarily considering paleoenvironmental information. On the basis of variables like continuity vs. discontinuity, regional histories may be contrasted. The rock-art archaeological evidence, as any other, may be then studied from this perspective. Even if it is difficult to assign chronologies to rock-art motifs, this is possible. The record of overlapping of motifs and techniques, together with the identification of different degrees of tonal series and/or varnish, provide a first approach (Re et al. 2005). Besides, as we have already emphasized, the convergence of different lines of evidence provides an integrative frame to assess these issues.

Archaeological cases

On the basis of published information, we compare in an exploratory frame data about the distribution of rock-art motifs in two regions from southern and northern Patagonia respectively (Figure 1). These evidences are referred to a biogeographical scheme with emphasis on cultural transmission processes and niche construction. The potential of rock-art evidences has been recently explored considering a geographical framework and proposing demographic inferences (Belardi and Goñi 2006; Fiore 2006a, 2006b; Muscio 2006; Re et al. 2009). We follow the distributional perspective, considering the artifact as the analytical unit (Foley 1981; Borrero 1993a); this is also valid for rock-art motifs (Aschero 1988).

Both regional cases show different degrees of advance. In the case from southern Patagonia, centered on the Strobel plateau, most of the data has been generated from a perspective that is similar to the one proposed here, highlighting the fertility of this theoretical approach. On the contrary, in northern Neuquén there is important information available, although it was generated from other perspectives, emphasizing regional culture-history issues (however, see Hajduk and Cúneo 2009, for a recent contribution emphasizing human geographic organization).

Strobel lake plateau

Strobel lake plateau (900 masl.) is placed in northwestern Santa Cruz province (Figure 3). A model that explains the distribution and diversity of rock-art engraved motifs in terms of the regional and supra-regional demographic dynamics during the late Holocene has been developed (Belardi and Goñi 2002, 2006; Goñi et al. 2007; Re et al. 2009). Differences have been noticed in terms of the distribution of engraved motifs in this plateau vs. painted motifs on Cardiel lake low altitude basin -placed right to the south of the plateau- which, together with technological evidence, were related to the differential use given to these spaces by hunter-gatherer populations. On the other hand, the high frequency and diversity of engraved motifs (geometrical and figuratives) on the Strobel plateau contrasts with other areas of central western Santa Cruz, where a lower density and diversity of designs for the late Holocene is found (Goñi et al. 2007; Re et al. 2009; see also Carden 2009). This has been the basis on which we propose the existence of population convergence on the Strobel plateau. Environmental conditions provide ecological support to the model: its location at the convergence of natural circulation paths, its seasonal pattern of use -concentrated in summer-, the availability of young guanacos as well as other fauna, and the availability of water and shelter provided by lagoons framed by basaltic walls. This matches with certain regional archaeological properties, like the equipment of space -shown by the high frequency of hunting blinds-, the preeminence of hunting activities also related to the high frequency of arrowheads and the emphasis on obsidian use as a raw material for lithic artifacts. It is worth mentioning that this raw material comes from the Pampa del Asador, located 40 km northwards (Espinosa and Goñi 1999; Stern 1999; Belardi et al. 2006).

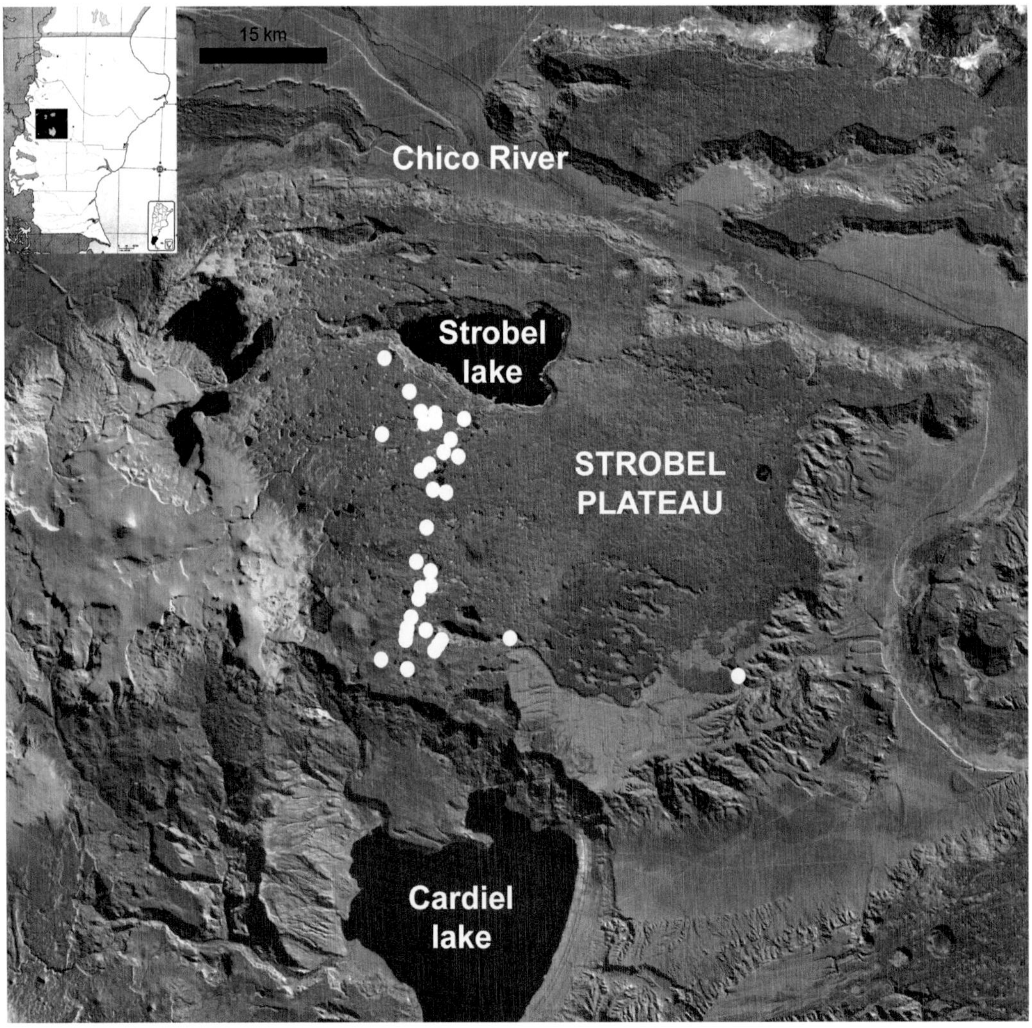

FIGURE 3. ARCHAEOLOGICAL SITES IN THE STROBEL PLATEAU.

The high frequency and diversity of motifs on the Strobel plateau may be interpreted as an evidence of communication, like a form of intentional transmission of information. In conjunction with other intentional modifications of the landscape, like the hunting blinds, and with unintended changes, like the surface concentrations of lithic materials, would have been part of the regional niche construction by human populations during the late Holocene.

The study of rock-art representations suggests different temporal instances in the occupation of the Strobel plateau (Figure 4). A first moment has been proposed which could have its initial stage during the mid-Holocene, characterized as an exploration stage. It is mainly represented by guanaco motifs with designs in which their bellies stand out, made by scraping and/or incision (Re 2009) (Figure 4). They have a differencial distribution, with a higher density on its southern part. During later times (at the late Holocene according to radiocarbonic datings), a larger amount and diversity of motifs is recorded (abstract: circles, halfcircles, lines; zoomorphic: guanacos, matuastos (lizard) and piches (armadillo) silhouettes, and feline, bird and guanaco tracks; and antropomorphic: complete figures, footprints and hands) (Figure 4). These are found widely and more or less homogeneously distributed across the plateau (Re and Guichon 2009) (Figures 3 and 4). According to varnish patterns and overlappings, temporal tendencies would also exist within this moment. It is for this time that we have proposed the population convergence hypothesis mentioned above (Belardi and Goñi 2006; Goñi et al. 2007), which may be understood as a definitive colonization of this space.

It has been suggested that during XVIth and XVIIth centuries, after the first contact with European populations, rock-art would still be performed, adding new designs to this repertoire (horse's silhouettes and tracks and human figures with crosses) (Re et al. 2009). Lastly, during the XVIIIth and XIXth centuries, rock-art would not be performed, due to changes in mobility and extensive use of the plateau and ultimately its subsequent abandonment (Re et al. 2009).

Given the tendencies identified on the basis of the analysis of rock-art representations, it is suggested that the occupations on the Strobel plateau would correspond to Cases 3 or 4 of Figure 2, as we have verified temporal changes in the use of space that may or may not be associated to demographic changes. However, it is difficult to evaluate this issue archaeologically on the basis of the information available so far.

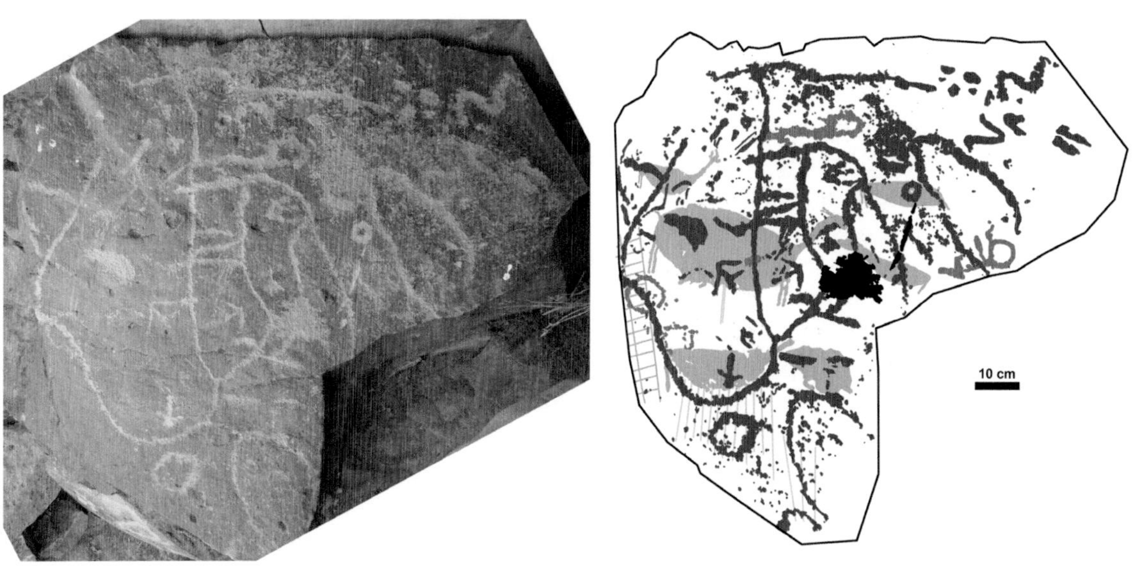

FIGURE 4. ROCK-ART MOTIFS FROM THE STROBEL PLATEAU.

Northern Neuquén

Neuquén province is located in the northern end of the Patagonia region (Figure 1). A clear prevalence of engravings emplaced in open-air contexts has been recorded (Schobinger 1956, 1962-1963; Fernández 1974-1976, 2000[1979]; Hajduk and Cúneo 2010). With certain variations, these engravings are also found in the southernmost part of Mendoza province, located right to the north (Gradin 1997-1998; Schobinger 2002). In the case of Neuquén, these engravings are associated with two main geographical features that run from north to south: the Cordillera del Viento and a segment of the Andes range. In fact, Hajduk and Cúneo (2010) suggest that the rock-art motifs are emplaced according with favored human circulation paths on a regional scale. Styllistically, the prevalence of abstract geometric motifs, assigned to the "paralelas" style by Menghin (1957), has been observed for this region (Fernández 2000[1979]) (Figure 5).

There are two patterns that become interesting here: firstly, the existence of a wide range of shared motifs in northern Neuquén and the Linares Range in Chile, for which the "Guaiquivilo" style has been defined (Niemeyer and Montané 1968; Fernández 2000 [1979]; Hajduk and Cúneo 2010). Second, when these different regions are related we can verify a situation that is comparable to that described by Belardi and Goñi (2006) for the Strobel plateau: south central Chile includes the motifs that are characteristic of northern Neuquén as well as a wide variety of motifs that are not present there, but that have been found in different regions of Patagonia (quantitative analysis in Fernández 2000[1979]: 69-74).

These observations of the authors mentioned are tentative and must be evaluated, even though they suggest a differential diversity of rock-art motifs among these adjacent regions. In a general level, the important aspect is that they may indicate an information circulation dynamics shared between spaces that have differences at the geographical and ecological level. A deeper analysis may contribute to identify geographical vectors of information circulation as well as temporal tendencies.

Conclusions

Both archaeological cases can be conceived as an instance of niche construction set up through the transmission of information, in which the information shared is loaded on different landscapes that are related at a supra-regional scale. In this geographical context of discussion, these spaces may be hierarchically arranged in terms of the place

they occupy in human circulation webs. It is interesting to note that both cases contrast heavily in terms of environmental variation. If looked at a wide spatial scale, central western Santa Cruz can be seen as ecologically similar; on the other hand, northern Neuquén and central Chile present a great climatic and ecological contrast, strengthening the potential of complementarity in a relatively small spatial dimension (given that these two regions are ruled by climatic systems that work in a decoupled fashion; Garreaud et al. 2009; discussion in Neme et al. 2005). As Whallon (2006) suggests, the spatial scale in which these contrasts are expressed may condition the level of cultural and biological interactions established.

The similarity described for the Strobel plateau and northern Neuquén cases may have interesting demographical implications that may be contrasted in the future. The Strobel plateau is interpreted as functionally associated to spaces found in lower places. For this reason, it would be a functionally specific (or logistic) place associated at a demographical level to neighbouring spaces. In the Neuquén-Chile case there is evidence that suggests, at least in a first instance, the existence of independent demographic nodes. We state that they would be comparable situations as regards the distribution of rock-art motifs, but that they would have a different background considering the distribution of demographic nodes. These conditions may suggest different information transmission processes dynamics.

FIGURE 5. ROCK-ART MOTIFS FROM HUENUL CAVE, NORTHERN NEUQUÉN (SCHOBINGER 1985; BARBERENA ET AL. 2010).

The issue that we stress here is the fertility of the theoretical frame proposed to relate independent lines of evidence that express themselves in different scales. In this way the different distributions of rock-art motifs may be compared with those of other artifactual designs; lithic raw materials with a known procedence and stable isotopes in human samples. These lines of evidence present different spatial and temporal resolution, and are related to different dimensions of geographical organization of human societies; for which reason their integration contributes to a deeper understanding of the archaeological landscape.

The different branches and implications of Darwin's thought have reached all academic disciplines (Lyman 2007). The great virtue is the integration of disciplines and research fields that at first sight seem to be different. We hope to have shown an example of the relationship betwen niche construction theory, cultural transmission and archaeological landscapes and thus show the importance of considering a biogeographical frame as a mid range instance that allows to connect the material record with certain postulates of the evolutionary theory. This approach to the archaeological record allows to consider cultural change in wide spatial scales, related now, not only with environment variations, but to consider the people themselves as modifiers of it. This theoretical perspective provides the confidence of a strong theoretical anchoring and a clear concern for the empirical adequation of ideas. In a few words, this is another great legacy of Charles Darwin.

Acknowledgments

We acnowledge Marcelo Cardillo and Hernán Muscio, coorganizers of the symposium where this paper was originally presented, for their comments. Research was carried out within the frame of projects granted by CONICET (PIP 6405, D 325/10), ANPCyT (26295) and the Universidad Nacional de la Patagonia Austral (UNPA-UARG A/183/2, A/213-1 and 29/A245-1). Project "Arqueología del Depto. Pehuenches, Provincia de Neuquén" also allow research in Northern Neuquén where we acknowledge the help provided by Estela Cúneo, Liliana Martínez and Pablo Azar.

References

Aschero, C. 1988. Pinturas rupestres, actividades y recursos naturales: un encuadre arqueológico. In: Arqueología Contemporánea Argentina. Actualidad y Perspectivas. Edited by: H. Yacobaccio, pp. 109-145. Ediciones Búsqueda, Buenos Aires.

Aunger, R. 2009. Human Communication as Niche Construction. In: Pattern and Process in Cultural Evolution. Edited by: S. Shennan, pp. 33-43. Berkeley, University of California Press.

Barberena, R. 2008. Arqueología y biogeografía humana en Patagonia meridional. Buenos Aires, Sociedad Argentina de Antropología.

Barberena, R., M.P., Pompei, C., Otaola, A., Gil, G., Neme, V. Durán and R. Hoguin. 2010. Pleistocene-Holocene Transition in Northern Patagonia: Evidences from Huenul Cave (Neuquén, Argentina). In press in Current Research in the Pleistocene.

Barrientos, G. 2009. El estudio arqueológico de la continuidad/discontinuidad biocultural: el caso del Sudeste de la Región Pampeana. In: Perspectivas Actuales en Arqueología Argentina. Edited by: R. Barberena, K. Borrazzo and L. A. Borrero, pp. 191-214. Buenos Aires, CONICET-IMHICIHU.

Belardi, J. B. 2005. Paisajes arqueológicos: un estudio comparativo de diferentes ambientes patagónicos. BAR International Series 1390. Oxford.

Belardi, J. B. and R. A. Goñi 2002. Distribución espacial de motivos rupestres en la cuenca del lago Cardiel (Patagonia Argentina). Boletín SIARB 16: 29-38.

2006. Representaciones rupestres y convergencia poblacional durante momentos tardíos en Santa Cruz (Patagonia argentina). El caso de la meseta del Strobel. In: Tramas en la Piedra. Producción y usos del arte rupestre. Edited by: D. Fiore and M. M. Podestá, pp. 85-94. Buenos Aires, WAC, SAA, AAINA.

Belardi, J. B., P. Tiberi, C. Stern and A. Súnico. 2006. Al este del cerro Pampa: ampliación del área de disponibilidad de obsidiana de Pampa del Asador (Provincia de Santa Cruz). Intersecciones en Antropología 7: 27-36.

Bettinger. R. 2008. Cultural Transmission and Archaeology. In: Cultural Transmission and Archaeology. Issues and Case Studies. Edited by: M. J. O'Brien, pp. 1-8. Washington, Society for American Archaeology.

Binford, L. R. 1983. Long Term Land Use. Patterns: Some Implications for Archaeology. In: Lulu Linear Punctated: Essays in Honor of George Irving Quimby. Edited by: R. C. Dunnell and D. K. Grayson, pp. 27-54, Anthropological Papers 72, Museum of Anthropology, University of Michigan, Ann Arbor.

Borrero. L. A. 1989-90. Evolución cultural divergente en la Patagonia austral. Anales del Instituto de la Patagonia (Serie Ciencias Sociales) 19: 133-139.

-1993a. Artefactos y evolución. Palimpsesto. Revista de Arqueología 3: 15-32.

-1993b. Site Formation Processes in Patagonia: Depositional rates and the properties of the Archaeological Record. In: Explotación de Recursos Faunísticos en Sistemas Adaptativos Americanos. Compiled by: J. L. Lanata. Arqueología Contemporánea 4: 107-121.

-2005. The Archaeology of the Patagonian Deserts: Hunter-Gatherers in a Cold Desert. In: Desert Peoples. Archaeological Perspectives. Edited by: P. Veth, M. Smith and P. Hiscock, pp.142-158. Oxford, Blackwell.

Borrero, L.A. and L.M. Manzi. 2007. Arqueología supra-regional y biogeografía en la Patagonia meridional. In: Arqueología de Fuego-Patagonia. Levantando piedras, desenterrando huesos… y develando arcanos. Edited by: F. Morello, A. Prieto, M. Martinic y G. Bahamonde, pp. 163-172. Punta Arenas, Ediciones CEQUA.

Briggs, J.C. and C.J. Humphries. 2004. Early Classics. In: Foundations of Biogeography. Edited by: M.V., Lomolino, D.F. Sax and J.H. Brown, pp. 5-13. Chicago, The University of Chicago Press.

Carden, N. 2009. Imágenes a través del tiempo. Arte rupestre y construcción social del paisaje en la Meseta Central de Santa Cruz. Buenos Aires, Sociedad Argentina de Antropología.

Darwin, C. 1859. On the Origins of Species by Means of Natural Selection; or the Preservation of Favoured Races in the Struggle for Life. Murray, London.

Eerkens, J. and C. Lipo. 2007. Cultural Transmission Theory and the Archaeological Record: Providing Context to Understanding Variation and Temporal Changes in Material Culture. Journal of Archaeological Research 15: 239-274.

Espinosa S. L. and R. A. Goñi 1999. ¡Viven!: una fuente de obsidiana en la provincia de Santa Cruz. In: Soplando en el viento... Actas de las Terceras Jornadas de Arqueología de la Patagonia, pp. 177-188. Edited by: Instituto Nacional de Antropología y Pensamiento Latinoamericano y Universidad Nacional del Comahue. Neuquén y Buenos Aires.

Fernández, J. 1974-1976. Estudios sobre el arte rupestre de la provincia del Neuquén. Anales de Arqueología y Etnología XXIX-XXXI: 5-36.

-2000 [1979]. Las piedras con marcas de la cordillera del Viento. Arte rupestre en el departamento Minas, Neuquén, Argentina. Buenos Aires, Publicaciones de la Sociedad Argentina de Antropología.

Fiore, D. 2006a. Poblamiento de imágenes: arte rupestre y colonización de la Patagonia. Variabilidad y ritmos de cambio en tiempo y espacio. In: Tramas en la piedra. Producción y usos del arte rupestre. Edited by: D. Fiore and M. M. Podestá, pp. 43-61. Buenos Aires, WAC, SAA, AAINA.
-2006b. Puentes de agua para el arte mobiliar: la distribución espacio-temporal de artefactos óseos decorados en Patagonia meridional y Tierra del fuego. Cazadores-recolectores del Cono Sur. Revista de Arqueología 1: 137-147.
Foley, R. 1981. Off-Site Archaeology and Human Adaptation in Eastern Africa. An Analysis of Regional Artefact Density in the Amboseli, Southern Kenia. Cambridge Monographs in African Archaeology 3. BAR International Series 97. Oxford.
Garreaud, R. D., M. Vuille, R., Compagnucci and J. Marengo. 2009. Present-day South American Climate. Palaeogeography, Palaeoclimatology, Palaeoecology 281 (3-4): 180-185.
Giller, P. S., A. A. Myers and B. R. Riddle. 2004. Earth History, Vicariance, and Dispersal. In: Foundations of Biogeography. Edited by: M.V. Lomolino, D. F. Sax and J. H. Brown. pp. 267-276. Chicago, The University of Chicago Press.
Goñi, R., J. B. Belardi, A. Re, A. Nuevo Delaunay, R. Molinari and L. Ferraro. 2007. Los grabados de la meseta del lago Strobel (Patagonia Argentina) desde una perspectiva regional. In: Actas del Primer Simposio Nacional de Arte Rupestre (Cusco, noviembre de 2004). Actes & Memoires de l´Ínstitut Francais d´Etudes Andines. Tomo 12: 427-438.
Gradin, C. J. 1997-1998. El arte rupestre del sur Mendocino entre los siglos VIII y XV de la Era. ¿Un área de conflicto o de convivencia? Relaciones de la Sociedad Argentina de Antropología XXII-XXIII: 7-23.
Hajduk, A. and E.M. Cúneo. 2009. Representaciones rupestres en la Cuenca del río Curi Leuvú (departamento Chos Malal, Provincia del Neuquén, República Argentina). Informe preliminar. In: Arqueología de la Patagonia. Una mirada desde el útimo confín. Edited by: M. Salemme, F. Santiago, M. Alvarez, E. Piana, M. Vázquez y E. Mansur, pp. 515-526. Editorial Utopías, Ushuaia,
Hiscock, P. 2008. Archaeology of Ancient Australia. Routledge, London-New York.
Lyman, R. L. 2003. The influence of time averaging and space averaging on the application of foraging theory in zooarchaeology. Journal of Archaeological Science 30: 595-610.
-2007. Archaeology's quest for a seat at the high table of anthropology. Journal of Anthropological Archaeology 26: 133-149.
Martínez, G. A. 2008-2009. Arqueología del curso inferior del río colorado: estado actual del conocimiento e implicaciones para la dinámica poblacional de cazadores-recolectores pampeano-patagonicos. In press in Cazadores recolectores del cono sur. Revista de arqueología 3.
Martínez, G. A. and Q. Mackie. 2003-2004. Late Holocene human occupation of the Quequén Grande River valley botton: settlement systems and an example of a built environment in the Argentine Pampas. Before Farming (1)1.
Muscio, H. J. 2006. Una aproximación evolutiva a la complejidad y al orden social durante el período temprano a través del estudio de representaciones rupestres de la quebrada de Matancillas (Puna argentina). Estudios Atacameños 31: 9-30.
-2009. Nicho y estrategia predominante. Dos conceptos necesarios en arqueología evolutiva. In: 150 años después… La vigencia de la teoría evolucionista de Charles Darwin. Edited by: M. C. Barboza, J. D. Avila, C. Píccoli y J. Conaglia Fernández, pp. 83-105. Universidad Nacional de Rosario, Facultad de Humanidades y Arte. Centro de Estudios Interdisciplinarios en Antropología.
Neme, G., A. F. Gil and V. Durán. 2005. Late Holocene in southern Mendoza (northwestern Patagonia): radiocarbon pattern and human occupation. Before Farming 2005 (2): 197-214.
Niemeyer, H. and J. Montané. 1968. El arte rupestre indígena en la zona Centro-Sur de Chile. In: Actas y Memorias del XXXVII Congreso Internacional de Americanistas, Volumen II, pp. 419-451. Buenos Aires.
O'Connell, J. 1995. Ethnoarchaeology needs a general theory of behavior. Journal of Archaeological Research 3(3): 205-255.
Odling-Smee, F. J., K. N. Laland and M. W. Feldman. 2003. Niche Construction. The Neglected Process in Evolution. Monographs in Population Biology 37. Princeton, Princeton University Press.
Pianka, E. 1982. Ecología Evolutiva. Ediciones Omega, Barcelona.
Re, A. 2009. Grabados de guanacos en la Patagonia austral: Abordaje a la representación del animal más importante para la dieta de los cazadores-recolectores. Paper presented to the Global Rock Art. Congreso Internacional de Arte Rupestre. Piauí, Brasil.

Re, A., J. B. Belardi and R. A. Goñi. 2009. Dinámica poblacional tardía en Patagonia meridional: su discusión y evaluación a través de la distribución de motivos rupestres. In: Crónicas sobre la Piedra. Arte rupestre de las Américas. Edited by: M. Sepulveda, L. Briones and J. Chacama, pp. 293- 309. Arica, Universidad de Tarapacá.

Re, A., R. Goñi, J. B. Belardi and F. Guichón. 2009. Evidencias de contacto en el arte rupestre de la Meseta del Strobel (Patagonia austral argentina). Paper presented to the Global Rock Art. Congreso Internacional de Arte Rupestre. Piauí, Brasil.

Re, A. and F. Guichón. 2009. Densidad y distribución de representaciones rupestres en la meseta del Strobel (Provincia de Santa Cruz). In: Arqueología de Patagonia: una mirada desde el último confín. Edited by: M. Salemme, F. Santiago, M. Álvarez, E. Piana, M. Vázquez and M.E. Mansur, Volume 1: 527-539. Ushuaia, Editorial Utopías.

Re, A., A. Nuevo Delaunay and L. Ferraro. 2005. Grabados en la meseta del lago Strobel (Provincia de Santa Cruz): el sitio Laguna del Faldeo Verde. Relaciones de la Sociedad Argentina de Antropología XXX: 245-256.

Richerson, P. J. and R. Boyd. 2005. Not by genes alone. How culture transformed human evolution. Chicago, University of Chicago Press.

Rindos, D. 1980. Symbiosis, instability and the origins and spread of agriculture: a new model. Current Anthropology 26: 751-772.

Rockman, M. and J. Steele. 2003. Colonization of Unfamiliar Landscapes. The Archaeology of Adaptation. Routledge, London-New York.

Scheinsohn, V. 2008. "Andando el carro se acomodan los zapallos": la perspectiva darwiniana y el registro arqueológico. In: Puentes hacia el pasado. Reflexiones teóricas en Arqueología. Edited by: D. Jackson S., Diego Salazar S. and A. Troncoso M., pp. 129-146. Santiago de Chile, Serie Monográfica de la Sociedad Chilena de Arqueología 1.

Schobinger, J. S. 1956. El arte rupestre de la Provincia del Neuquén. Anales de Arqueología y Etnología XXII: 115-227.

-1962-1963. Nuevos petroglifos de la Provincia del Neuquén. Anales de Arqueología y Etnología XXVII-XXVIII: 151-171.

-1985. Áreas intermedias o marginales. In: Cazadores de la Patagonia y agricultores andinos. Arte rupestre de la Argentina. Edited by: J. Schobinger and C.J. Gradin, pp. 80-91. Madrid, Ediciones Encuentro.

Stern, C. R.

1999. Black obsidian from central-south Patagonia; chemical characteristics, sources and regional distribution of artifacts. In: Soplando en el viento. Actas de las III Jornadas de Arqueología de la Patagonia, pp. 221234. Edited by: Instituto Nacional de Antropología y Pensamiento Latinoamericano y Universidad Nacional del Comahue. Neuquén y Buenos Aires.

Taçon, P. 2008. Marks of possession: the archaeology of territory and cross-cultural encounter in Australia and South Africa. In: Handbook of Landscape Archaeology. Edited by: B. David and J. Thomas, pp. 218-227. World Archeological Congress. Research Handbooks in Archaeology . Series Editors: G. Nicholas and J. Hollowell. Walnut Creek.

Thomas, D. H. 1983. The archaeology of Monitor Valley. 1. Epistemology. In: Anthropological Papers of the American Museum of Natural History, 58 (1). New York.

Veth, P. M. 1993. Islands in the Interior. The Dynamics of Prehistoric Adaptations within the Arid Zone of Australia. Ann Arbor, International Monographs in Prehistory, Archaeological Series 3.

Whallon, R. 2006. Social networks and information: Non-"utilitarian" mobility among hunter-gatherers. Journal of Anthropological Archaeology 25 (2): 259-270.

Wiessner, P.

1983. Style and Social Information in Kalahari San Projectile Points. American Antiquity 48 (2):253-276.

Wobst, M. 1977. Stylistic Behavior and Information Exchange. En: Papers for the Director: Research Essays in Honor of James B. Griffin. Editado por C. E. Cleland. University of Michigan. Museum of Anthropology, Anthropological Papers, 61:317-342. Ann Arbor, Michigan.

Yacobaccio, H. D. 1988. Introducción. En: Arqueología Contemporánea Argentina. Actualidad y Perspectivas. Editado por H. Yacobaccio, pp. 7-12. Ediciones Búsqueda, Buenos Aires.

Zeanah, D. W. and S. R. Simms. 1999. Modeling the Gastric: Great Basin Subsistence Studies since 1982 and the Evolution of General Theory. In: Models for the Millenium. Great Basin Anthropology Today. Edited by: C. Beck, pp. 118-139. Salt Lake City, The University of Utah Press.